Carpentry and Joinery

NVQ and Technical Certificate **Level 3**

www.heinemann.co.uk

✓ Free online support
✓ Useful weblinks
✓ 24 hour online ordering

0845 630 44 44

Heinemann

Part of Pearson

Heinemann is an imprint of Pearson Education Limited, a company incorporated in England and Wales, having its registered office: Halley Court, Jordan Hill, Oxford OX2 8EJ. Registered company number: 3099304

www.heinemann.co.uk

Heinemann is a registered trademark of Pearson Education Limited

Text © Carillion Construction Ltd 2007, 2009

First published 2007

12 11 10 09
10 9 8 7 6 5 4 3 2

British Library Cataloguing in Publication Data is available from the British Library on request.

ISBN 978 0 435 32578 7

Edited by Sarah Christopher
Designed by HL Studios
Typeset by HL Studios
Printed in Italy by Rotolito

Illustrated by HL studios
Original illustrations © Pearson Education Limited 2007, 2009

Cover design by GD Associates
Cover photo © Pearson Education/Ben Nicholson

Every effort has been made to contact copyright holders of material reproduced in this book. Any omissions will be rectified in subsequent printings if notice is given to the publishers.

Websites

The websites used in this book were correct and up to date at the time of publication. It is essential for tutors to preview each website before using it in class so as to ensure that the URL is still accurate, relevant and appropriate. We suggest that tutors bookmark useful websites and consider enabling students to access them through the school/college intranet.

Contents

Acknowledgements

The authors and publisher would like to thank the following individuals and organisations for permission to reproduce photographs:

Alamy / Adrian Sherratt p63; Alamy / Andrew Paterson p186; Alamy / AT Willett p164; Alamy / David J Green p17; Alamy / David R. Frazier Photolibrary, Inc p191 (right); Alamy / Eric Nathan p86; Alamy / ImageBroker pp307, 323; Alamy / Justin Kase pp7, 141, 166; Alamy / Kris Mercer p153; Alamy / Libby Welch p146; Alamy / Nic Hamilton p5; Alamy / PintailPIctures pp83, 87; Alamy / PlainPictures GmbH & Co. KG p16; Alamy / The Photolibrary Wales p21; Alamy / Trish Grant p285; Construction Photography pp84, 119, 120 (bottom left); Construction Photography / DIY Photolibrary p190; Construction Photography / Buildpix pp137, 163; Construction Photography / Chris Henderson p9; Construction Photography / CP Stock p51; Construction Photography / Darren Holden p1; Construction Photography / David Stewart Smith p191 (right); Construction Photography / DIY Photolibrary pp33, 160; Construction Photography / Paul McMullin pp49, 191 (left), 257; Construction Photography / Xavier de Canto p25; Corbis p223; Corbis / Creasource p31; Corbis / Helen King p203; Corbis / Holger Scheibe / Zefa p210; Corbis / Martin Meyer / Zefa p37; Creatas p138; Getty Images / PhotoDisc pp14, 139; Jupiter Images / Photos.com p121 (bottom); Pearson Education / Gareth Boden pp3, 109 (left and right) 112, 113 (all), 114, 120 (top and bottom right), 121 (top left), 159, 186 (all except step 5), 187 (all), 211 (top), 230 (all), 308, 309 (both), 317 (right), 318; Pearson Education / Jules Selmes pp72, 173, 186 (step 5), 191 (left), 192 (both), 193 (all), 194 (both), 208, 209 (both), 211 (bottom two), 212 (both), 266, 267 (both), 268, 281, 288, 292; Philip Parkhouse p145; Photographers Direct / Bjorn Beheydt p28; Photographers Direct / David Griffiths pp26, 121 (top right); Science Photo Library / Garry Watson p8 (left); Science Photo Library / Scott Camazie p16; Shutterstock / David Hughes p170; Shutterstock / David Lee p148; Shutterstock / Stavklem p167; Startrite pp289, 297, 301, 305; Toolbank p317 (left); Topham Picturepoint p8 (middle and right)

Introduction

This book has been written based on a concept used in Carillion Training Centres for many years. That concept is about providing learners with the necessary information they need to support their studies and at the same time ensuring it is presented in a style which they find both manageable and relevant.

The contents of this book have been put together by Kevin Jarvis who has a wealth of knowledge and experience in both training for NVQ and technical certificates in his trade.

It builds upon material covered in *Carpentry and Joinery NVQ and Technical Certificate Level 2*, which introduced readers to the principles behind many of the carpentry topics that will be explored in greater depth here.

This book has been produced to help the learner build a sound knowledge and understanding of all aspects of the NVQ and Technical Certificate requirements associated with their trade. It has also been designed to provide assistance when revising for Technical Certificate end tests and NVQ job knowledge tests.

Each chapter of this book relates closely to a particular unit of the NVQ or Technical Certificate and aims to provide just the right level of information needed to form the required knowledge and understanding of that subject area.

This book provides a brief introduction to the supervisory role that a Level 3 student will be working towards as well as providing the knowledge on the tools, materials and methods of work allowing you to complete work activities safely, effectively and productively. Upon completion of your studies, this book will remain a valuable source of information and support when carrying out your work activities.

For further information on how the contents of this book matches to the unit requirements of the NVQ and Advanced Construction Award, please use the detailed mapping document which can be found on our website www.heinemann.co.uk

How this book can help you

You will discover the following features throughout this book, each of which has been designed and written to increase and improve your knowledge and understanding.

- **Photographs** – identify a tool or material or will help you to follow a step-by-step procedure.

- **Illustrations** – give you more information about a concept or procedure.

- **Definition** – new or difficult words are picked out in bold in the text and defined in the margin.

- **Remember** – highlight key concepts or facts.

- **Safety tip** – guidance to help you work safely.

- **Did you know?** – interesting facts about the building trade.

- **Find out** – short activities designed to lead you to find further information and gain better understanding of a topic area.

- **Activity** – small tasks designed to test your understanding.

- **FAQs** – frequently asked questions together with informative answers from the experts.

- **Knowledge refresher** – questions to test your knowledge and recall of a topic.

- **What would you do?** – real-life situations designed to make you think about what you would do. (Answers can be found on the Tutor Resource Disk that accompanies this book.)

- **Glossary** – contains definitions of all the **bold** words and phrases found in the text. A great quick reference tool.

chapter 1

Health and safety

OVERVIEW

The construction industry is the largest industry in the UK – and the most dangerous. In the past 25 years, almost 3000 people have died from injuries received carrying out construction work; many more have been seriously injured or made ill.

Thankfully, there has been a slight reduction over the last few years: according to the Health and Safety Executive (HSE), 79 people died in 2006/07 compared to 72 people in 2007/08. This minor reduction is due to increased regulations and improved training and awareness. Health and safety training is now a major part of any apprenticeship.

Level 2 gave a good grounding in health and safety, and informed you of what you need to know and do. This Level 3 book will refresh the main points and give you a more comprehensive insight into health and safety in the construction industry.

This chapter will cover:

- health and safety legislation
- health and welfare
- hazards, emergencies and accidents
- risk assessments.

Health and safety legislation

While at work, whatever your location or type of work, you need to be aware that there is important **legislation** you must comply with. Health and safety legislation is there not just to protect you, but also states what you must and must not do to ensure that no workers are placed in a situation **hazardous** to themselves or others. Each piece of legislation covers your own responsibilities as an employee and those of your employer – it is vital that you are aware of both.

What is legislation?

Legislation means a law or set of laws passed by Parliament, often called an Act. There are hundreds of Acts covering all manner of work from hairdressing to construction. Each Act states the duties of the **employer** and **employee**. If an employer or employee does something they shouldn't – or doesn't do something they should – they can end up in court and be fined or even imprisoned.

Approved Code of Practice (ACoP), guidance notes and safety policies

As well as Acts, there are two sorts of codes of practice and guidance notes: those produced by the **Health and Safety Executive (HSE)**, and those created by companies themselves. Most large construction companies – and many smaller ones – have their own guidance notes, which go further than health and safety law. For example, the law states that that everyone must wear safety boots in a hazardous area, but a company's code may state that everyone must wear safety boots at all times. This is called taking a **proactive** approach, rather than a **reactive** one.

Most companies have some form of **safety policy** outlining the company's commitment and stating what they plan to do to ensure that all work is carried out as safely as possible. As an employee, you should make sure you understand the company's safety policy as well as their codes of practice. If you act against company policy you may not be prosecuted in court, but you could still be disciplined by the company or even fired.

If you are acting as a supervisor, you will need to ensure that staff you are supervising understand the safety policy too. The safety policy may even require you to take on further responsibilities as a supervisor, such as running safety drills and checks: if so, you will need to understand what is involved in these and ask for any necessary support or training from your employer.

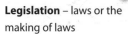

Definition

Legislation – laws or the making of laws

Hazardous – dangerous or unsafe

Definition

Health and Safety Executive (HSE) – government organisation that enforces health and safety law in the UK

Proactive – acting in advance, before something happens (such as an accident)

Reactive – acting after something happens, in response to it

Find out

Health and safety regulations are constantly being updated and amended. Log on to the HSE website (www.hse.gov.uk/construction) to see what recent updates there have been to the different regulations.

Health and safety legislation you need to be aware of

There are some 20 pieces of legislation you will need to be aware of, each of which sets out requirements for employers and often employees. One phrase often comes up here – *'so far as is reasonably practicable'*. This means that health and safety must be adhered to at all times, but must take a common sense, practical approach.

For example, the Health and Safety at Work Act 1974 (HASAWA) states that an employer must *so far as is reasonably practicable* ensure that a safe place of work is provided. Yet employers are not expected to do everything they can to protect their staff from lightning strikes, as there is only a 1 in 800,000 chance of this occurring – this would not be reasonable!

We will now look at the regulations that will affect you most.

The Health and Safety at Work Act 1974

The Health and Safety at Work Act was first introduced on 31 July 1974. It is described as follows:

> ' … an act to make further provisions for securing the health, safety and welfare of persons at work, for protecting others against risks to health and safety in connection with the activities of persons at work, for controlling the keeping and use and preventing the unlawful acquisition, possession and use of dangerous substances, and for controlling certain emissions into the atmosphere; to make further provision with respect to the employment medical advisory services; to amend the law relating to building regulations, and the Building (Scotland) Act 1959; and for connected purposes.'

This is what the law states, but what does it mean to your employer or to you as an employee?

The HASAWA applies to all types and places of work and to employers, employees, the self-employed, sub-contractors and even suppliers. The act is there to protect not only the people at work but also the general public, who may be affected in some way by the work that has been or will be carried out.

The main **objectives** of the health and safety at work act are to:

- ensure the health, safety and welfare of all persons at work

- protect the general public from all work activities

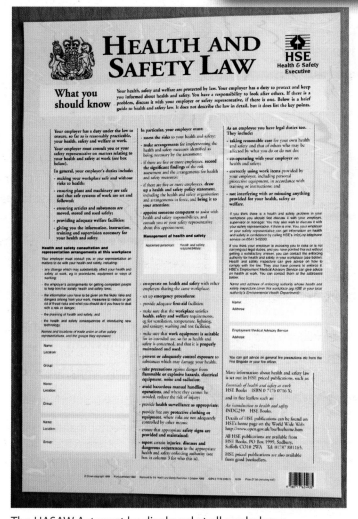

The HASAW Act must be displayed at all workplaces

- control the use, handling, storage and transportation of explosives and highly flammable substances
- control the release of noxious or offensive substances into the atmosphere.

To ensure that these objectives are met there are duties for all employers, employees and suppliers.

Employer's duties

Employers must:

- provide safe **access** and **egress** to and within the work area
- provide and maintain a safe working environment
- provide and maintain plant and machinery that is safe and without risks to health
- provide information, instruction, training and supervision to ensure the health and safety at work of all employees
- ensure safety and the absence of risks to health in connection with the handling, storage and transportation of articles and substances
- have a written safety policy that must be revised and updated regularly, and ensure all employees are aware of it
- involve trade union safety representatives, where appointed, in all matters relating to health and safety
- provide and not charge for **personal protective equipment (PPE)**.

Employee's duties

The employee must:

- take reasonable care for his/her own health and safety
- take reasonable care for the health and safety of anyone who may be affected by his/her acts or **omissions**
- co-operate with his/her employer or any other person to ensure legal **obligations** are met
- not misuse or interfere with anything provided for their health and safety
- use any equipment and safeguards provided by his/her employer.

Those supervising others will need to make sure that staff are using all safety equipment in the appropriate way, and are taking any necessary steps to ensure that they do not jeopardise their own or others' safety.

Employees cannot be charged for anything that has been done or provided for them to ensure that legal requirements on health and safety are met. The self-employed and sub-contractors have the same duties as employees – and if they have employees of their own, they must obey the duties set down for employers.

Definition

Access – entrance, a way in

Egress – exit, a way out

PPE – personal protective equipment, such as gloves, a safety harness or goggles

---

Definition

Omission – something that has not been done or has been missed out

Obligation – something you have a duty or a responsibility to do

---

Supplier's duties

Persons designing, manufacturing, importing or supplying articles or substances for use at work must ensure that:

- articles are designed and constructed so that they will be safe and without risk to health at all times while they are being used or constructed

- substances will be safe and without risk to health at all times when being used, handled, transported and stored

- tests on articles and substances are carried out as necessary

- adequate information is provided about the use, handling, transporting and storing of articles or substances.

HASAWA, like most of the other acts mentioned, is enforced by the Health and Safety Executive (HSE). HSE inspectors visit sites and have the power to:

- enter any premises at any reasonable time

- take a police constable with them

- examine and investigate anything on the premises

- take samples

- take possession of any dangerous article or substance

- issue improvement notices giving a company a certain amount of time to sort out a health and safety problem

- issue a **prohibition** notice stopping all work until the site is deemed safe

- **prosecute** people who break the law including employers, employees, self-employed, manufacturers and suppliers.

Activity

The HSE produces a variety of posters, which they use as a way of conveying information and as a training tool. Look at the employee's duties under the HASAWA and design a poster that can be used to convey the message of one or more of the employee's duties.

Provision and Use of Work Equipment Regulations 1998 (PUWER)

These regulations cover all new or existing work equipment – leased, hired or second-hand. They apply in most working environments where the HASAWA applies, including all industrial, offshore and service operations.

PUWER covers starting, stopping, regular use, transport, repair, modification, servicing and cleaning.

Legislation is there to protect employees and the public alike

'Work equipment' includes any machinery, appliance, apparatus or tool, and any assembly of components that are used in non-domestic premises. Dumper trucks, circular saws, ladders, overhead projectors and chisels would all be included, but substances, private cars and structural items all fall outside this definition.

The general duties of the regulations require equipment to be:

- suitable for its intended purpose and only to be used in suitable conditions
- maintained in an efficient state and maintenance records kept
- used, repaired and maintained only by a suitably trained person, when that equipment poses a particular risk
- able to be isolated from all its sources of energy
- constructed or adapted to ensure that maintenance can be carried out without risks to health and safety
- fitted with warnings or warning devices as appropriate.

In addition, the regulations require:

- all personnel to be trained and deemed competent before using any work equipment
- all those who use, supervise or manage work equipment to be suitably trained
- access to any dangerous parts of the machinery to be prevented or controlled
- injury to be prevented from any work equipment that may have a very high or low temperature
- suitable controls to be provided for starting and stopping the work equipment
- suitable emergency stopping systems and braking systems to be fitted to ensure the work equipment is brought to a safe condition as soon as reasonably practicable
- suitable and sufficient lighting to be provided for operating the work equipment.

Control of Substances Hazardous to Health Regulations 2002

The Control of Substances Hazardous to Health Regulations 2002 (COSHH) state how employees and employers should work with, handle, store, transport and dispose of potentially hazardous substances (substances that might negatively affect your health) including:

- substances used directly in work activities (e.g. adhesives or paints)
- substances generated during work activities (e.g. dust from sanding wood)
- naturally occurring substances (e.g. sand dust)
- biological agents (e.g. bacteria).

Correct labelling and storage of hazardous substances saves lives

These substances can be found in nearly all work environments. All are covered by COSHH regulations except asbestos and lead paint, which have their own regulations.

To comply with COSHH regulations, eight steps must be followed:

Step 1 Assess the risks to health from hazardous substances used or created by your activities.

Step 2 Decide what precautions are needed.

Step 3 Prevent employees from being exposed to any hazardous substances. If prevention is impossible, the risk must be adequately controlled.

Step 4 Ensure control methods are used and maintained properly.

Step 5 Monitor the exposure of employees to hazardous substances.

Step 6 Carry out health **surveillance** to ascertain if any health problems are occurring.

Step 7 Prepare plans and procedures to deal with accidents such as spillages.

Step 8 Ensure all employees are properly informed, trained and supervised.

Safety tip

Not all substances are labelled, and sometimes the label may not match the contents. If you are in any doubt, do not use or touch the substance.

Identifying a substance that may fall under the COSHH regulations is not always easy, but you can ask the supplier or manufacturer for a COSHH data sheet, outlining the risks involved with it. Most substance containers carry a warning sign stating whether the contents are corrosive, harmful, toxic or bad for the environment.

Corrosive

Toxic hazard

Risk of explosion

Common safety signs for corrosive, toxic and harmful materials

The Personal Protective Equipment at Work Regulations 1992

These regulations, known as PPER, cover all types of PPE, from gloves to breathing apparatus. After doing a risk assessment and once the potential hazards are known, suitable types of PPE can be selected. PPE should be checked prior to issue by a trained and competent person and in line with the manufacturer's instructions. Where required, the employer must provide PPE free of charge along with a suitable and secure place to store it.

The employer must ensure that the employee knows:

- the risks the PPE will avoid or reduce
- its purpose and use
- how to maintain and look after it
- its limitations.

The employee must:

- ensure that they are trained in the use of the PPE prior to use
- use it in line with the employer's instructions
- return it to storage after use
- take care of it, and report any loss or defect to their employer.

Remember

PPE must only be used as a last line of defence.

The Control of Noise at Work Regulations 2005

At some point in your career in construction, you are likely to work in a noisy working environment. These regulations help protect you against the consequences of being exposed to high levels of noise, which can lead to permanent hearing damage.

Damage to hearing has a range of causes, from ear infections to loud noises, but the regulations deal mainly with the latter. Hearing loss can result from one very loud noise lasting only a few seconds, or from relatively loud noise lasting for hours, such as a drill.

The regulations state that the employer must:

- assess the risks to the employee from noise at work

- take action to reduce the noise exposure that produces these risks

Noise doesn't have to be loud to damage your hearing

- provide employees with hearing protection or, if this is impossible, reduce the risk by other methods

- make sure the legal limits on noise exposure are not exceeded

- provide employees with information, instruction and training

- carry out health surveillance where there is a risk to health.

Anyone who is supervising the work of others may be asked to (or may wish to) monitor noise levels in the work area and advise the employer if any problems arise from excessive noise exposure.

The Work at Height Regulations 2005

In these regulations, 'working at height' covers not only working from scaffolding or other high places, but also working at ground level or below if a person can be injured by falling from that level.

Falls are the biggest cause of death and serious injury in the construction industry, so these regulations make sure that employers do all that they can to reduce the risk of injury or death from working at height.

Did you know?

Noise is measured in **decibels (dB)**. The average person may notice a rise of 3dB, but with every 3dB rise, the noise is doubled. What may seem like a small rise is actually very significant.

The employer's main duty is to do all that is reasonably practicable to prevent anyone falling. In doing this, the employer must:

- avoid work at height where possible
- use any equipment or safeguards that will prevent falls
- use equipment and any other methods that will minimise the distance and consequences of a fall.

The employer must also ensure that:

- all work at height is properly planned
- weather conditions that can endanger health and safety are taken into account
- those involved in the work are trained and competent
- the place where the work is being done is safe
- equipment is appropriately inspected
- the risks from fragile surfaces are properly controlled
- the risks from falling objects are properly controlled.

As an employee, you must follow any training given to you, report any hazards to your supervisor and use any safety equipment made available to you.

The Electricity at Work Regulations 1989

These regulations cover any work involving the use of electricity or electrical equipment. An employer has the duty to ensure that the electrical systems their employees come into contact with are safe and regularly maintained. They must also have done everything the law states to reduce the risk of their employees coming into contact with live electrical currents.

The Manual Handling Operations Regulations 1992

More than a third of all over-three-day injuries reported to the HSE are caused by manual handling – the transporting or supporting of loads by hand or bodily force. These regulations cover all work activities in which a person, rather than a machine, does the lifting.

The regulations require employers to:

- **avoid** the need for manual handling, so far is as reasonably practicable
- **assess** the risk of injury from any hazardous manual handling that cannot be avoided
- **reduce** the risk of injury from hazardous manual handling, so far is as is reasonably practicable.

The best way for an employer to meet these regulations is to carry out manual handling risk assessments. In a risk assessment, there are four considerations:

- **Load** – is it heavy, sharp-edged, difficult to hold?

- **Individual** – is the individual small, pregnant, in need of training?

- **Task** – does the task require holding goods away from the body, or repetitive twisting?

- **Environment** – is the floor uneven, are there stairs, is it raining?

After the assessment, the situation must be monitored constantly and updated or changed if necessary.

The Reporting of Injuries, Diseases and Dangerous Occurrences Regulations 1995 (RIDDOR)

Under the Reporting of Injuries, Diseases and Dangerous Occurrences Regulations 1995 (RIDDOR), employers have a duty to report work-related deaths, major injuries, over-three-day injuries, work-related diseases and dangerous occurrences (near miss accidents).

Reporting accidents and ill health at work is a legal requirement. The information reported to the HSE enables it to identify where and how risks arise and to investigate serious accidents, so that advice can be given to prevent any reoccurrences.

Reporting any relevant injury, disease or dangerous occurrence covered by RIDDOR can be done by telephone, online at the HSE website, by e-mail or by post.

Other acts to be aware of

You should also be aware of the following legislation:

- The Fire Precautions (Workplace) Regulations 1997

- The Fire Precautions Act 1991

- The Highly Flammable Liquids and Liquid Petroleum Gases Regulations 1972

- The Lifting Operations and Lifting Equipment Regulations 1998

- The Construction (Health, Safety and Welfare) Regulations 1996

- The Environmental Protection Act 1990

- The Confined Spaces Regulations 1997

- The Working Time Regulations 1998

- The Health and Safety (First Aid) Regulations 1981

- The Construction (Design and Management) Regulations 1994.

You can find out more at the library or online.

Find out

Look into the other regulations listed here via the government website www.hse.gov.uk

Activity

Think of a simple task within your occupation. What regulations would apply to that task? Write down these regulations, then state what your responsibilities would be under each regulation.

Knowledge refresher

1 What is legislation?

2 What is an approved code of practice?

3 What is the purpose of a safety policy?

4 What does 'so far is as reasonably practicable' mean?

5 Who enforces the health and safety regulation?

6 State three main objectives of the HASAW Act 1974.

7 State the main objective of the Working at Height Regulations.

8 State four duties included in the PUWER regulations.

9 How can an occurrence covered by RIDDOR be reported?

What would you do?

1 You are about to carry out a task that involves chemicals. The regulations state that risk assessments should be carried out before starting the task. The risk assessments show that expensive measures will need to be put into place if you are to complete the task – and this would mean that you would not make a decent profit. Cheaper options are available but will not offer the same protection. What should you do? What outcomes could there be? What could have been done before getting to this stage?

2 Your employer has asked you and a colleague to work on a scaffold that has not been checked and you think is a bit unsafe. Your colleague, who seems to be on good terms with the boss, says there is no problem, but you raise your concerns with your employer. He ridicules you and replies, 'If you won't do it, I'll find someone who will.' What should you do? What could be the repercussions? What rights do you have as an employee? What obligations does your employer have?

Health and welfare

As a worker in the construction industry, you will be at constant risk unless you adopt a good health and safety attitude. By following the rules and regulations, and by taking reasonable care of yourself and others – especially where acting as a supervisor – you will become a safe worker and reduce the chance of injuries and accidents. Given the statistics on safety, the supervisor's role is crucial here: few other people will be a better position to understand the day-to-day work of a site, be in touch with those doing the labour and spot 'danger points' where accidents or ill health could occur.

The two most common risks to a construction worker

Before reading further, what do you think the two most common risks to a construction worker might be? Think about the industry you are working in and the hazards and risks that exist.

In fact, the most common health and safety risks a construction worker faces are:

- accidents
- ill health.

Accidents

We often hear the saying 'accidents will happen', but in the construction industry, the truth is that most accidents are caused by human error – someone does something they shouldn't or, just as importantly, does not do something they should. Accidents often happen when someone is hurrying, not paying attention, trying to cut corners or costs, or has not received the correct training.

If an accident happens, you or the person it happened to may be lucky enough to escape uninjured. More often, an accident will result in an injury, whether minor (e.g. a cut or a bruise), major (e.g. loss of a limb) or even fatal. The most common causes of fatal accidents in the construction industry are:

- falling from scaffolding
- being hit by falling objects and materials
- falling through fragile roofs
- being hit by forklifts or lorries
- cuts
- infections
- burns
- **electrocution**.

A fall could be fatal so a safety harness should always be worn

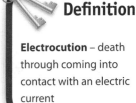

Definition

Electrocution – death through coming into contact with an electric current

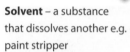

Remember

Everyone has a responsibility for health and safety, but accidents and health problems still happen too often. Make sure you do what you can to prevent them.

Definition

Solvent – a substance that dissolves another e.g. paint stripper

Vibration white finger – a condition that can be caused by using vibrating machinery (usually for a very long period of time). The blood supply to the fingers is reduced which causes pain, tingling and sometimes spasms (shaking)

Remember

Some health problems do not show symptoms straight away – and what you do now can affect you greatly in later life.

Ill health

In the construction industry, you will be exposed to substances or situations that may be harmful to your health. Some of these health risks may not be noticeable straight away and it may take years for symptoms to be noticed and recognised.

Ill health can result from:

- exposure to dust (or asbestos fibres) – breathing problems and cancer
- exposure to **solvents** or chemicals – dermatitis and other skin problems
- lifting heavy or difficult loads – back injury and pulled muscles
- exposure to loud noise – hearing problems and deafness
- exposure to sunlight, which can cause skin cancer
- using vibrating tools – **vibration white finger** and other hand problems.

Substance abuse

'Substance abuse' is a general term that covers drinking alcohol and taking drugs, as well as other substances.

Drinking alcohol is dangerous at work; going to the pub for lunch and having just one drink can lead to slower reflexes and reduced concentration.

Taking drugs or inhaling solvents at work is not only illegal, but is also highly dangerous to you and everyone around you, as reduced concentration can lead to accidents.

Although not a form of abuse as such, taking drugs prescribed by your doctor or even over-the-counter painkillers can be dangerous. Many of these medicines carry warnings such as 'may cause drowsiness' or 'do not operate heavy machinery'. It is better to be safe than sorry, so always ensure you follow any instructions on prescriptions and, if you feel drowsy or unsteady, stop work immediately.

Staying healthy

As well as watching for hazards, you must also look after yourself and stay healthy.

One of the easiest ways to do this is to wash your hands regularly: this prevents hazardous substances entering your body through ingestion (swallowing). You should always wash your hands after going to the toilet and before eating or drinking.

Other precautions that you can take include the following:

- ensuring that you wear barrier cream to protect yourself from the sun
- using the correct PPE to ensure that your back, arms and legs are sufficiently covered
- drinking only water that is labelled as drinking water.

Welfare facilities

Welfare facilities are things that an employer must provide to ensure a safe and healthy workplace.

Always wash your hands to prevent ingesting hazardous substances

- **Toilets** – The number provided depends on how many people are intending to use them. Males and females can use the same toilet providing the door can be locked from the inside. Toilets should ideally be flushable with water or, if this is not possible, with chemicals.

- **Washing facilities** – Employers must provide a basin large enough for people to wash their hands, face and forearms, with hot and cold running water, soap and a way to dry your hands. Showers may be needed if the work is very dirty or if workers are exposed to toxic or corrosive substances.

- **Drinking water** – A supply of clean drinking water should be available, from a mains-linked tap or bottled water. Mains-linked taps need to be clearly labelled as drinking water; bottled drinking water must be stored where there is no chance of contamination.

- **Storage or dry room** – Every building site must have an area where workers can store clothes not worn on site, such as coats and motorcycle helmets. If this area is to be used as a drying room, adequate heating must be provided.

- **Lunch area** – Every site must have facilities for taking breaks and lunch well away from the work area. There must be shelter from the wind and rain, with heating as required, along with tables and chairs, a kettle or urn and a means of heating food.

Hazards, emergencies and accidents

The building industry can be a very dangerous place to work and there are certain hazards that all workers need to be aware of.

The main types of hazards that you will face are:

- falling from height

- tripping

- chemical spills

- burns

- electrical hazards

- fires.

Falling from height and fires have been covered in detail in Level 2, so here we will look at the remaining hazards.

Tripping

The main cause of tripping is poor housekeeping. Whether working on scaffolding or on ground level, an untidy workplace is an accident waiting to happen. All workplaces should be kept tidy and free of debris. All off-cuts should be put either straight into a skip or, if you are not near a skip, in a wheelbarrow. Not only will this prevent trip hazards, it will also prevent costly clean-up operations at the end of the job and will promote a good professional image.

An untidy work site can present many trip hazards

Chemical spills

A chemical spill can be anything from a minor inconvenience to a major disaster. Most spills are small and create minimal or no risk. If the material involved is not hazardous, it can be cleaned up by normal operations such as brushing or mopping. However, on some occasions the spill may be on a larger scale and may involve a hazardous material. It is important to know what to do before the spill happens so that remedial action can be prompt, and harmful effects minimised.

Before a hazardous substance is used, a COSHH or risk assessment will have been made, which should include a plan for dealing with a spill. This, in turn, should mean that the materials required for dealing with the spill should be readily available.

Burns

Burns can occur not only from the obvious source of fire and heat but also from materials such as cement or painter's solvents. Even electricity can cause burns. It is vital when working with any material that you are aware of the hazards it may present and that you take the necessary precautions.

If you receive a burn, or find yourself having to help a colleague who has one, you must act quickly and carefully. Get in touch with a First Aider straight away, and then contact the emergency services.

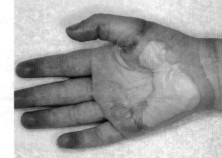

Fire, heat, chemicals and electricity can cause burns

Electrical hazards

Electricity is a killer. According to the HSE, around 30 workers a year die from electricity-related accidents, with over 1000 more being seriously injured.

One of the main problems with electricity is that it is invisible. You do not even have to be working with an electric tool to be electrocuted: working too close to live overhead cables, plastering a wall with electric sockets, carrying out maintenance work on a floor, or drilling into a wall can all lead to an electric shock.

Electric shocks may not always be fatal: electricity can also cause burns, muscular problems and cardiac (heart) problems. Despite the common perception, the level of voltage is not a direct guide to the level of injury or danger of death: a small shock from static electricity may contain thousands of volts, but has very little current behind it. However, it is generally true that the lower the voltage, the lower the chance of death occurring.

Household wiring

There are two main types of voltage in use in the UK. These are 230 V and 110 V. The standard UK power supply is 230 V – this is what all the sockets in your house are.

Contained within the wiring there should be three wires: the live and neutral, which carry the alternating current, and the earth wire, which acts as a safety device. The three wires are colour-coded:

- live – brown
- neutral – blue
- earth – yellow and green.

These colours comply with current European colours. Some older properties you work on may have the following colours:

- live – red
- neutral – black
- earth – yellow and green.

230 V has been deemed unsafe on construction sites, so 110 V must be used here. 110 V, identified by a yellow cable and different style of plug, works from a transformer which converts the 230 V to 110 V.

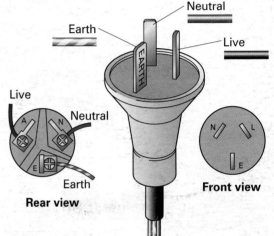

Figure 1.1 Colour coding of the wires in a 110 V plug

Dealing with electric shocks

In helping a victim of an electric shock, the first thing you must do is disconnect the power supply, if it is safe to do and will not take long to find – touching the power source may put you in danger.

If the victim is in contact with something portable, such as a drill, try to move it away using a non-conductive object, such as a wooden broom. Time is precious and separating the victim from the source can prove an effective way to speed the process. Do not attempt to touch the affected person until they are free and clear of the supplied power, and don't touch the victim until you are sure the power supply is turned off. Be especially careful in wet areas, such as bathrooms, since water can conduct electricity, and electrocuting yourself is also possible.

A 110 V plug

People 'hung up' in a live current flow may think they are calling out for help but it is likely that no sound will be heard from them. When the muscles contract under household current (most electrocutions happen from house current at home), the person affected will seem to be in a 'locked up' state, unable to move or react to you.

Using a wooden object, swiftly and strongly knock the person free, trying not to injure them, and land them clear of the source. You may lift or remove the item, if it is safe for you to do so, with the same wooden item. This is not recommended on voltages that exceed 500 V. Do not attempt any of this unless you are wearing shoes with rubber or some form of insulated sole: bare or socked feet will allow the current to flow to the ground through your body as well.

First aid procedures for an electric shock victim

- Check if you are alone. If there are other people around, instruct them to call an ambulance right away.

- Check the victim's breathing, and see if they can respond to you.

- If the area is safe for you to be in, and you have removed the object or have cut off its power supply, shout to the victim to see if they are conscious. At this stage, do not touch them.

- Check once again to see if the area is safe. If you are satisfied that it is safe, start resuscitating the victim if required. If you have no first aid knowledge, call emergency services for an ambulance.

Emergencies

So far, we have covered most of the emergencies that occur on site, such as accidents and fires, but there are other emergencies that you need to be aware of, such as security alerts and bomb scares.

At your site induction, it should be made perfectly clear to you what you should do in the event of an emergency. You also should be made aware of any sirens or warning noises that accompany each and every type of emergency, such as bomb scares or fire alarms. Some sites may have variations on sirens or emergency procedures, so it is vital that you pay attention and listen to all instructions. If you are unsure, always ask.

Knowledge refresher

1   What are the two most common risks to construction workers?

2   List three things that can cause ill health, and what health problems they can create.

3   State five welfare facilities that must be available.

4   State the correct colour for 110 V cables/plugs.

5   How should a chemical spill be dealt with?

6   What colour(s) are earth wires in standard 230 V plugs?

What would you do?

You are inducted onto a new site, but the person hosting the induction does not give very clear instructions. The induction finishes and you are still unsure of what the site rules are or what do to in case of an emergency. What should/would you do?

Accidents

Reporting accidents

When an accident occurs, there are certain things that must be done. All accidents need to be reported and recorded in the accident book, and the injured person must report to a trained first aider to receive treatment. Serious accidents must be reported under RIDDOR: these regulations state that your employer must report to the HSE any accident that results in:

- death

- major injury

- an injury that means the injured person is not at work for more than three consecutive days.

Activity

Working in a group, create a presentation to highlight a type of accident or injury and open a discussion to consider ways of avoiding the accident.

Report of an Accident, Dangerous Occurrence or Near Miss

Date of incident _____ Time of incident _12:00_____

Location of incident _Toilets_____

Details of person involved in accident

Name _Hamzs_____ Date of birth _26/03/98___ Sex _F/M_
                                                      Not Sure

Address _____

_____ Occupation _escort_____

Date off work (if applicable) _____ Date returning to work _1 year from the_

Nature of injury _____rectum damage_____

Management of injury   ☐ First Aid only   ☑ Advised to see doctor
                        ☑ Sent to casualty ☑ Admitted to hospital

Account of accident, dangerous occurrence or near miss
(Continued on separate sheet if necessary)

Hamzs was bammed senseless by
an African American and was
then given a pirate (torched in
the shen the got spank in his eye.

Witnesses to the incident
(Names, addresses and occupations)

Donal heard screaming in the
toilets.

Was the injured person wearing PPE? If yes, what PPE? _If by PPE_
you mean condom then Yes

Signature of person completing form _____

Occupation _____ Date _____

Figure 1.2 A typical accident book page

The accident book

The accident book is completed by the person who had the accident or, if this is not possible, by someone representing the injured person. The accident book will ask for some basic details about the accident, including:

- who was involved
- what happened
- where it happened
- the day and time of the accident
- details of any witnesses to the accident
- the address of the injured person
- what PPE was being worn
- what first aid treatment was given.

As well as reporting accidents, 'near misses' must also be reported. This is because near misses are often the accidents of the future: reporting near misses might identify a problem and can prevent accidents from happening. In this way, a company can try to prevent future accidents, rather than just dealing with the ones that happen.

First aid

In the unfortunate event of an accident on site, first aid may have to be administered. If there are more than five people on a site, a qualified first aider must be present at all times. On large building sites, there must be several first aiders. During your site induction, you will be made aware of who the first aiders are and where the first aid points are situated. A first aid point must have the relevant first aid equipment to deal with the types of injuries that are likely to occur. However, first aid is only the first step and, in the case of major injuries, the emergency services should be called.

First aid box

A good first aid box should have plasters, bandages, antiseptic wipes, latex gloves, eye patches, slings, wound dressings and safety pins. Other equipment, such as eye wash stations, must also be available if the work being carried out requires it.

Risk assessments

You will have noticed that most of the legislation we have looked at requires risk assessments to be carried out. The Management of Health and Safety at Work Regulations 1999 require every employer to make suitable and sufficient assessment of:

- the risks to the health and safety of his/her employees to which they are exposed while at work

- the risks to the health and safety of persons not in his/her employment arising out of or in connection with his/her work activities.

As a Level 3 candidate, it is vital that you know how to carry out a risk assessment. Often you may be in a position where you are given direct responsibility for this, and the care and attention you take over it may have a direct impact on the safety of others. You must be aware of the dangers or hazards of any task, and know what can be done to prevent or reduce the risk.

Even an everyday task like cutting the grass has its own dangers

There are five steps in a risk assessment – here we use cutting the grass as an example:

Step 1 Identify the hazards

When cutting the grass the main hazards are from the blades or cutting the wire, electrocution and any stones that may be thrown up.

Step 2 Identify who will be at risk

The main person at risk is the user but passers-by may be struck by flying debris.

Step 3 Calculate the risk from the hazard against the likelihood of an accident taking place

The risks from the hazard are quite high: the blade or wire can remove a finger, electrocution can kill and the flying debris can blind or even kill. The likelihood of an accident happening is

Did you know?

We all carry out risk assessments hundreds of times a day. For example, every time we boil a kettle, we do a risk assessment without even thinking about it: for example, by checking the kettle isn't too full, or the cable frayed, and by keeping children out of the way.

Definition

Making a risk assessment – measuring the dangers of an activity against the likelihood of accidents taking place

Activity

Using a fictitious task that you are familiar with, create a risk assessment, showing what the risks are and what measures you could introduce to control them.

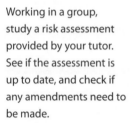

Definition

RCD – residual current device, a device that will shut the power down on a piece of electrical equipment if it detects a change in the current, thus preventing electrocution

Activity

Working in a group, study a risk assessment provided by your tutor. See if the assessment is up to date, and check if any amendments need to be made.

medium: you are unlikely to cut yourself on the blades, but the chance of cutting through the cable is medium, and the chance of hitting a stone high.

Step 4 Introduce measures to reduce the risk

Training can reduce the risks of cutting yourself; training and the use of an **RCD** can reduce the risk of electrocution; and raking the lawn first can reduce the risk of sending up stones.

Step 5 Monitor the risk

Constantly changing factors mean any risk assessment may have to be modified or even changed completely. In our example, one such factor could be rain.

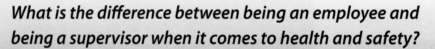

FAQ

What is the difference between being an employee and being a supervisor when it comes to health and safety?

As a supervisor, you need to be aware of the different pieces of legislation and consider them when you are overseeing other people's work. You may be given responsibility to help your employer comply with their part of legislation too.

How do I find out what safety legislation is relevant to my job?

Ask your employer or contact the HSE at www.hse.gov.uk.

How do I find my company's safety policy?

Ask your supervisor or employer.

When do I need to do a risk assessment?

A risk assessment should be carried out if there is any chance of an accident happening. To be on the safe side, you should make a risk assessment before starting each task.

Do I need to read and understand every regulation?

No. It is part of your employer's duty to ensure that you are aware of what you need to know.

Knowledge refresher

1 Who should fill in an accident report form?

2 List five pieces of information that should be entered onto an accident report form.

3 Why is it important to report 'near misses'?

4 Name five items that it's a good idea to have in a first aid kit.

5 Briefly explain what a risk assessment is.

6 State why risk assessments should be monitored.

What would you do?

1 You are a self-employed sub-contractor working on a job when you accidentally cut your finger. The cut is not too deep but may require medical treatment. You put a plaster on it and get back to work, but later on you notice that the bleeding hasn't really stopped and you start to feel light-headed. If you stop work, you won't get paid and could lose out on future work. What do you do? What could the repercussions be for your work? What could the repercussions be for your health?

2 You are asked to do a job and on getting there you notice that the risk assessments have already been done. You quickly scan through them and notice that they are not done for this job but have been based on a previous job. You speak to the supervisor and he states that the job is similar, so the risks should be the same. What do you think?

chapter 2

Building documentation

OVERVIEW

In the construction industry you come across a wide range of documentation, and as a Level 3 apprentice you will encounter different types of documents more frequently. This chapter covers the main building documentation you will see, explaining what each type of documentation is and what it is used for.

The types of documentation covered in this chapter are:

- contract documents
- the Building Regulations
- general site paperwork.

Contract documents

Activity

Think of a task you are familiar with, such as building a wall or constructing a roof, and write a specification for that task. Remember to include things that would not be shown on a drawing, but would be needed to complete a task.

Contract documents are vital to a construction project. They are created by a team of specialists – the architect, structural engineer, services engineer and quantity surveyor – who first look at the draft of drawings from the architect and client. Just which contract documents this team goes on to produce will vary depending on the size and type of work being done, but will usually include:

- plans and drawings
- specification
- schedules
- bill of quantities
- conditions of contract.

Plans and drawings are covered in Chapter 4, so here we will start with the specifications.

Specification

A good 'spec' helps avoid confusion when dealing with sub-contractors or suppliers

The specification or 'spec' is a document produced alongside the plans and drawings and is used to show information that cannot be shown on the drawings. Specifications are almost always used, except in the case of very small contracts. A specification should contain:

- **site description** – a brief description of the site including the address
- **restrictions** – what restrictions apply such as working hours or limited access
- **services** – what services are available, what services need to be connected and what type of connection should be used
- **materials description** – including type, sizes, quality, moisture content, etc.
- **workmanship** – including methods of fixing, quality of work and finish.

The specification may also name sub-contractors or suppliers, or give details such as how the site should be cleared, and so on.

Schedules

A schedule is used to record repeated design information that applies to a range of components or fittings. Schedules are mainly used on bigger sites where there are multiples of several types of house (4-bedroom, 3-bedroom, 3-bedroom with dormers, etc.), each type having different components and fittings. The schedule avoids the wrong component or fitting being put in the wrong house. Schedules can also be used on smaller jobs such as a block of flats with 200 windows, where there are six different types of window.

The need for a specification depends on the complexity of the job and the number of repeated designs that there are. Schedules are mainly used to record repeated design information for:

- doors
- windows
- ironmongery
- joinery fitments
- sanitary components
- heating components and radiators
- kitchens.

A schedule is usually used in conjunction with a range drawing and a floor plan.

The following are basic examples of these documents, using a window as an example:

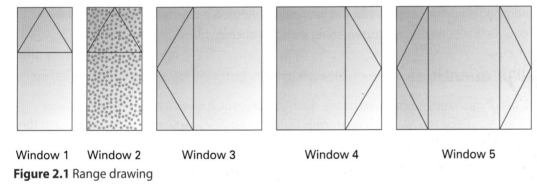

Window 1   Window 2   Window 3   Window 4   Window 5

Figure 2.1 Range drawing

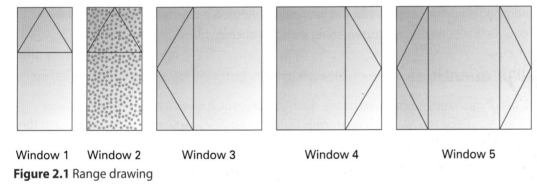

Figure 2.2 Floor plan

WINDOW SCHEDULE		
WINDOW	LOCATIONS	NOTES
Window 1	Stairwell	
Window 2	Bathroom En-suite	Obscure glass
Window 3	Bedroom 1 Bedroom 2	
Window 4	Bedroom 3 Master bedroom	
Window 5	Bedroom 4	

Figure 2.3 Schedule for a window

The schedule shows that there are five types of window, each differing in size and appearance; the range drawing shows what each type of window looks like; and the floor plan shows which window goes where. For example, the bathroom window is a type two window, which is 1200 × 600 × 50 mm with a top-opening sash and obscure glass.

Bill of quantities

The bill of quantities is produced by the quantity surveyor. It gives a complete description of everything that is required to do the job, including labour, materials and any items or components, drawing on information from the drawings, specification and schedule. The same single bill of quantities is sent out to all **prospective** contractors so they can submit a tender based on the same information – this helps the client select the best contractor for the job.

Activity

Bills of quantities are used to help contractors provide a tender for a contract. Think of a simple task, then create a bill of quantities for that task, including labour, materials, and so on.

Every item needed should be listed on the bill of quantities

All bills of quantities contain the following information:

- **preliminaries** – general information such as the names of the client and architect, details of the work and descriptions of the site

- **preambles** – similar to the specification, outlining the quality and description of materials and workmanship, etc.

- **measured quantities** – a description of how each task or material is measured with measurements in metres (linear and square), hours, litres, kilograms or simply the number of components required

- **provisional quantities** – approximate amounts where items or components cannot be measured accurately

- **cost** – the amount of money that will be charged per unit of quantity.

The bill of quantities may also contain:

- any costs that may result from using sub-contractors or specialists

- a sum of money for work that has not been finally detailed

- a sum of money to cover contingencies for unforeseen work.

This is an extract from a bill of quantities that might be sent to prospective contractors, who would then complete the cost section and return it as their tender.

Item ref No	Description	Quantity	Unit	Rate £	Cost £
A1	Treated 50 × 225 mm sawn carcass	200	M		
A2	Treated 75 × 225 mm sawn carcass	50	M		
B1	50 mm galvanised steel joist hangers	20	N/A		
B2	75 mm galvanised steel joist hangers	7	N/A		
C1	Supply and fit the above floor joists as described in the preambles				

Figure 2.4 Sample extract from a bill of quantities

To ensure that all contractors interpret and understand the bill of quantities consistently, the Royal Institution of Chartered Surveyors and the Building Employers' Confederation produce a document called the *Standard Method of Measurement of Building Works* (SMM). This provides a uniform basis for measuring building work, for example stating that carcassing timber is measured by the metre whereas plasterboard is measured in square metres.

Conditions of contract

Almost all building work is carried out under a contract. A small job with a single client (e.g. a loft conversion) will have a basic contract stating that the contractor will do the work to the client's satisfaction, and that the client will pay the contractor the agreed sum of money once the work is finished.

Larger contracts with clients such as the Government will have additional clauses, terms or **stipulations** written into the contract. Most large contracts are awarded to companies not solely on the basis of cost, but also taking into account the benefits of the package offered for both the community and the environment.

Clauses, terms or stipulations may include any of the following.

Environmental management

At the tendering stage of a project, most government bodies or large, privately run companies will ask potential contractors to specify how they will manage waste and recycling, and how they will minimise the impact on the environment (for example, by re-planting uprooted trees or moving plants and vegetation instead of destroying them).

Companies will be monitored against their own targets and, if they fail to meet them, may face a financial penalty. Because of this, tendering companies may underestimate their targets for a tender: for example, by stating that they will use 15 per cent recycled/re-used materials on the project rather than 20 per cent.

Health and safety management

On particularly large contracts, tendering companies will provide written safety policy and mission statements, stating how they will do the job safely.

Local workforce/suppliers/materials management

A good tender will specify the proportion of local workers that will be used during the work, in an effort to help bolster the local economy and reduce local unemployment. Using local suppliers or materials that are produced locally is also considered good practice for successful tenders.

Community regeneration management

Some companies will commit to upgrading or improving communities as part of their tender. For example, a company tendering to build a block of flats may commit to building play parks, youth clubs or other community amenities.

After the tender has been agreed, further conditions may be added to the contract, as follows.

Variations

A variation is a modification of the original drawing or specification. The architect or client must give the contractor written confirmation of the variation, then the contractor submits a price for the variation to the quantity surveyor (or client, on a small job). Once the price is accepted, the variation work can be completed.

Interim payment

An **interim** payment schedule may be written into the contract, meaning that the client pays for the work in instalments. The client may pay an amount each month, linked to how far the job has progressed, or may make regular payments regardless of how far the job has progressed.

Final payment

Here the client makes a one-off full payment once the job has been completed to the specification. A final payment scheme may also have additional clauses included, such as:

- **retention**
  This is when the client holds a small percentage of the full payment back for a specified period (usually six months). It may take some time for any defects to show, such as cracks in plaster. If the contractor fixes the defects, they will receive the retention payment; if they don't fix them, the retention payment can be used to hire another contractor to do so.

- **penalty clause**
  This is usually introduced in contracts with a tight deadline, where the building must be finished and ready to operate on time. If the project overruns, the client will be unable to trade in the premises and will lose money, so the contractor will have to compensate the client for lost revenue.

It is vital you check the exact terms of each contract

> ### Did you know?
>
> On a poorly run contract, a penalty clause can be very costly and could incur a substantial payment. In an extreme case, the contractor may end up making a loss instead of a profit on the project.

Knowledge refresher

1   Name four people who are involved in creating contract documents.

2   What is the purpose of a specification?

3   List four things that should be contained in a specification.

4   What is the purpose of a schedule?

5   Who produces the bill of quantities and what is its main purpose?

6   State why companies will add stipulations to their tender.

7   Describe what a variation is in regards to contract conditions.

8   Describe a penalty clause.

What would you do?

You have been invited to tender a bid for a large public contract. Business has been slow, and you really need it if you are to keep your business afloat and avoid redundancies.

Two of the other tenders concern you. One is priced so low that, if you match it, you may make a small loss. In the other, the contractor promises to recycle 45% of materials, to use only sustainable materials and to employ 70% of the local workforce – matching this may mean you have to lay off some workers and may only make a small profit.

What should you do? What stipulations could you introduce to help improve your bid? What could the consequences be of not getting the contract – or, indeed, of getting it?

The Building Regulations

The Building Regulations were first introduced in the late 19th century to improve the appalling housing conditions common then. The Public Health Act 1875 allowed local authorities to make their own laws regarding the planning and construction of buildings. There were many grey areas and **inconsistencies** between local authorities, especially where one authority bordered another.

This system remained in place for almost a century until the Building Regulations 1965 came into force. These replaced all local laws with a uniform Act for all in England and Wales to follow. The only exception was inner London, which was covered by the London Building Acts. The Government passed a new law in 1984, setting up the Building Regulations 1985 to cover all England and Wales, including inner London.

The current law is the Building Regulations 2000, amended in April 2006 to take into account things such as wheelchair access and more environmentally friendly practices. The current law also covers all England and Wales.

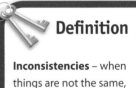

Definition

Inconsistencies – when things are not the same, not consistent

Building Regulations help protect the environment

Find out

For more about the Building Regulations, visit www.ukbuildingstandards.org.uk

Remember

Building regulations can change over time. Always be sure you are using the most updated version.

Scotland is governed slightly differently and is covered by the Building (Scotland) Act 2003. Northern Ireland is covered by the Building (Amendment) Regulations (Northern Ireland) 2006 which came into effect on November 2006.

The main purpose of the Building Regulations is to ensure the health, safety and welfare of all people in and around buildings as well as to further energy conservation and to protect the environment. The regulations apply to most new buildings as well as any alterations to existing buildings, whether they are domestic, commercial or industrial. Many projects also require planning permission, which will be covered in Chapter 3.

The regulations are broken down into several categories:

- Part A – Structural safety
- Part B – Fire safety
- Part C – Resistance to moisture and weather
- Part D – Toxic substances
- Part E – Resistance to sound
- Part F – Ventilation
- Part G – Hygiene
- Part H – Drainage and waste disposal
- Part J – Heat-producing appliances
- Part K – Protection from falling
- Part L – Conservation of fuel and power
- Part M – Access to and use of buildings
- Part N – Glazing safety
- Part P – Electrical safety.

Each of these sections contains an 'approved document', detailing what is covered by that part of the regulations:

Approved document A

A1 – Loading

A2 – Ground movement

A3 – Disproportionate collapse

Approved document B

B1 – Means of warning and escape

B2 – Internal fire spread (linings)

B3 – Internal fire spread (structure)

B4 – External fire spread

B5 – Access and facilities for the fire service

Approved document C

C1 – Site preparation and resistance to contaminates

C2 – Resistance to moisture

Approved document D

D1 – Cavity insulation

Approved document E

E1 – Protection against sound from other parts of the building and adjoining buildings

E2 – Protection against sound within a dwelling-house, etc.

E3 – Reverberation in the common internal parts of buildings containing flats or rooms for residential purposes

E4 – Acoustic conditions in schools

Approved document F deals only with ventilation

Approved document G

G1 – Sanitary conveniences and washing facilities

G2 – Bathrooms

G3 – Hot water storage

Approved document H

H1 – Foul water drainage

H2 – Wastewater treatment systems and cesspools

H3 – Rainwater drainage

H4 – Building over sewers

H5 – Separate systems of drainage

H6 – Solid waste storage

Approved document J

J1 – Air supply

J2 – Discharge of products of combustion

J3 – Protection of building

J4 – Provision of information

J5 – Protection of liquid fuel storage systems

J6 – Protection against pollution

Approved document K

K1 – Stairs, ladders and ramps

K2 – Protection from falling

K3 – Vehicle barriers and loading bays

K4 – Protection from collision with open windows, skylights and ventilators

K5 – Protection against impact from and trapping by doors

Approved document L

L1A – Conservation of fuel and power in new dwellings

L1B – Conservation of fuel and power in existing dwellings

L2A – Conservation of fuel and power in new buildings other than dwellings

L2B – Conservation of fuel and power in existing buildings other than dwellings

Approved document M

M1 – Access and use

M2 – Access to extensions to buildings other than dwellings

M3 – Sanitary conveniences in extensions to buildings other than dwellings

M4 – Sanitary conveniences in dwellings

Approved document N

N1 – Protection against impact

N2 – Manifestation of glazing

N3 – Safe opening and closing of windows, skylights and ventilators

N4 – Safe access for cleaning windows, etc.

Approved document P

P1 – Design and installation of electrical installations

These are the types of work classified as needing Building Regulations approval:

- the erection of an extension or building
- the installation or extension of a service or fitting which is controlled under the regulations
- an alteration project involving work which will temporarily or permanently affect the ongoing compliance of the building, service, or fitting with the requirements relating to structure, fire, or access to and the use of the building
- the insertion of insulation into a cavity wall
- the underpinning of the foundations of a building
- work affecting the thermal elements, energy status or energy performance of the building.

If you are unsure whether the work you are going to carry out needs Building Regulations approval, contact the local authority.

The Building Regulations are enforced by two types of building control bodies: local authority building control and Approved Inspector building control. If you wish to apply for approval, you must contact one of these bodies.

If you use an Approved Inspector, you must contact the local authority to tell them what is being done where, stating that the Inspector will be responsible for the control of the work.

If you choose to go to the local authority, there are three ways of applying for consent:

- **Full plans** – Plans are submitted to the local authority along with any specifications and other contract documents. The local authority scrutinises these and makes a decision.
- **Building notice** – A less detailed amount of information is submitted (but more can be requested) and no decision is made. The approval process is determined by the stage the work is at.
- **Regularisation** – This is a means of applying for approval for work that has already been completed without approval.

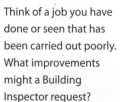
The Building Inspector will make regular visits to ensure that the work is being carried out to the standards set down in the application, and that no extra unapproved work is being done. Often the contractor will tell the Inspector when the job has reached a certain stage, so that they can come in and check what has been done. If the Inspector is not informed at key stages, he/she can ask for the work to be opened up to be checked.

Building Regulations approval is not always given but there is an appeals procedure. For more information, contact your local authority.

The Building Inspector will need to be involved at every key stage

Knowledge refresher

1 What is the main purpose of the Building Regulations?

2 Which approved document deals with stairs?

3 Which approved document deals with conservation of fuel and power?

4 What does approved document F deal with?

5 List four types of work that would require Building Regulations approval.

6 Who can you contact to check if the work you are doing requires Building Regulations approval?

7 What is the role of the Building Inspector?

What would you do?

You are part way through building an extension when the client asks for an alteration to the original plans. You think that this alteration may need Building Regulations approval, but applying now would put the job back a few weeks, and you are already under time pressure. The client says they do not care about the Building Regulations; they want the work done now, or they will stop paying you. What should you do? What could the repercussions of your actions be?

General site paperwork

No building site could function properly without a certain amount of paperwork. Here is a brief, but not exhaustive, description of some of the other documents you may encounter. Some companies will have their own forms to cover such things as scaffolding checks.

Timesheet

Timesheets record hours worked, and are completed by every employee individually. Some timesheets are basic, asking just for a brief description of the work done each hour, but some can be complicated. In some cases timesheets may be used to work out how many hours the client will be charged for.

P. Gresford Building Contractors

Timesheet

Employee _____ **Project/site** _____

Date	Job no.	Start time	Finish time	Total time	Travel time	Expenses
M						
Tu						
W						
Th						
F						
Sa						
Su						
Totals						

Employee's signature _____

Supervisor's signature _____

Date _____

Figure 2.5 Timesheet

Day worksheets

Day worksheets are often confused with timesheets, but are different as they are used when there is no price or estimate for the work, to enable the contractor to charge for the work. Day worksheets record work done, hours worked and sometimes materials used. They are also used when variation orders or extra work is added to a contract.

P. Gresford Building Contractors

Day worksheet

Customer _Chris MacFarlane_   Date _____

Description of work being carried out _____

Hang internal door in kitchen.

Labour	Craft	Hours	Gross rate	TOTALS
Materials	**Quantity**	**Rate**	**% addition**	
Plant	**Hours**	**Rate**	**% addition**	

Comments

Signed _____   Date _____

Site manager/foreman signature _____

Figure 2.6 Day worksheet

P. Gresford Building Contractors

Job sheet

Customer   Chris MacFarlane

Address   1 High Street
                Any Town
                Any County

Work to be carried out

Hang internal door in kitchen

Special conditions/instructions

Fit with door closer
3 × 75 mm butt hinges

Figure 2.7 Job sheet

Job sheet

A job sheet is similar to a day worksheet – it records work done – but is used when the work has already been priced. Job sheets enable the worker to see what needs to be done and the site agent or working foreman to see what has been completed.

VARIATION TO PROPOSED WORKS AT 123 A STREET

REFERENCE NO:

DATE _____

FROM _____

TO _____

POSSIBLE VARIATIONS TO WORK AT 123 A STREET

ADDITIONS
OMISSIONS

SIGNED -------------------------------------

Variation order

This sheet is used by the architect to make any changes to the original plans, including omissions, alterations and extra work.

Figure 2.8 Variation order

CONFIRMATION FOR VARIATION TO PROPOSED WORKS AT 123 A STREET

REFERENCE NO:

DATE _____

FROM _____

TO _____

I CONFIRM THAT I HAVE RECEIVED WRITTEN INSTRUCTIONS
FROM _____
POSITION _____
TO CARRY OUT THE FOLLOWING POSSIBLE VARIATIONS TO THE ABOVE NAMED CONTRACT

ADDITIONS
OMISSIONS

SIGNED -------------------------------------

Confirmation notice

This is a sheet given to the contractor to confirm any changes made in the variation order, so that the contractor can go ahead and carry out the work.

Figure 2.9 Confirmation notice

Orders/requisitions

A requisition form or order is used to order materials or components from a supplier.

Activity

Ordering the wrong size, type or amount of materials can cause delays or cost a lot of money on a project. Think of a simple task and write a materials requisition for that task.

P. Gresford Building Contractors

Requisition form

Supplier _____   Order no. _____

_____

_____   Serial no. _____

Tel no. _____   Contact _____

Fax no. _____   Our ref _____

Contract/Delivery address/Invoice address   Statements/applications for payments to be sent to

_____

_____   _____

Tel no. _____   _____

Fax no. _____   _____

Item no.	Quantity	Unit	Description	Unit price	Amount

Total £ _____

Payment terms _____   Date _____

Originated by _____

Authorised by _____

Figure 2.10 Requisition form

Delivery notes

Definition

Discrepancies – when there is a difference or variation between two things that should be the same

Delivery notes are given to the contractor by the supplier, and list all the materials and components being delivered. Each delivery note should be checked for accuracy against the order (to ensure what is being delivered is what was asked for) and against the delivery itself (to make sure that the delivery matches the delivery note). If there are any **discrepancies** or if the delivery is of a poor quality or damaged, you must write on the delivery note what is wrong before signing it and ensure the site agent is informed so that he/she can rectify the problem.

Bailey & Sons Ltd

Building materials supplier

Tel: 01234 567890

Your ref: AB00671

Our ref: CT020   **Date:** 17 Jul 2006

Order no: 67440387

Invoice address: **Delivery address:**
Carillion Training Centre,   Same as invoice
Deptford Terrace, Sunderland

Description of goods	Quantity	Catalogue no.
OPC 25kg	10	OPC1.1

Comments:

Date and time of receiving goods:

Name of recipient (caps):

Signature:

Figure 2.11 Delivery note

INVOICE   **JARVIS BUILDING SUPPLIES**
                              3RD AVENUE
                              THOMASTOWN

L Weeks Builders
4th Grove
Thomastown

Quantity	Description	Unit price	Vat rate	Total
30	Galvanised joist hangers	£1.32	17.5%	£46.53
			TOTAL	£46.53

To be paid within 30 days from receipt of this invoice

Please direct any queries to 01234 56789

Figure 2.12 Invoice

Invoices

Invoices come from a variety of sources such as suppliers or sub-contractors, and state what has been provided and how much the contractor will be charged for it.

Remember

Invoices may need paying by a certain date – fines for late payment can sometimes be incurred – so it is important that they are passed on to the finance office or financial controller promptly.

JARVIS BUILDING SUPPLIES
3RD AVENUE
THOMASTOWN

Customer ref_____

Customer order date_____

Delivery date_____

Item no	Qty Supplied	Qty to follow	Description	Unit price
1	30	0	Galvanised joist hangers	£1.32

Delivered to: L Weeks Builders
4th Grove
Thomastown
Customer signature ---------------------------

Delivery records

Delivery records list all deliveries over a certain period (usually a month), and are sent to the contractor's Head Office so that payment can be made.

Figure 2.13 Delivery record

Remember

Remember – you should always check a delivery note against the order and the delivery itself, then write any discrepancies or problems on the delivery note *before* signing it.

Daily report/site diary

This is used to pass general information (deliveries, attendance, etc.) on to a company's Head Office.

Figure 2.14 Daily report or site diary

DAILY REPORT/SITE DIARY

PROJECT_____
DATE_____

Identify any of the following factors, which are affecting or may affect the daily work activities and give a brief description in the box provided

WEATHER ( )   ACCESS ( )   ACCIDENTS ( )   SERVICES ( )
DELIVERIES ( )   SUPPLIES ( )   LABOUR ( )   OTHER ( )

SIGNED ---
POSITION ---

Accident and near miss reports

It is a legal requirement that a company has an accident book, in which reports of all accidents must be made. Reports must also be made when an accident nearly happened, but did not in the end occur – known as a 'near miss'. It is everyone's responsibility to complete the accident book. If you are also in a supervisory position you will have the responsibility to ensure all requirements for accident reporting are met.

Safety tip

If you are involved in or witness an accident or near miss, make sure it is entered in the book – for your own safety and that of others on the site. If you don't report it, it's more likely to happen again.

Report of an Accident, Dangerous Occurrence or Near Miss

Date of incident _____   **Time of incident** _____

Location of incident _____

Details of person involved in accident

Name _____   Date of birth _____

Address _____

_____

_____   Occupation _____

Date off work (if applicable) _____   **Date returning to work** _____

Nature of injury _____

Management of injury   ☐ First Aid only   ☐ Advised to see doctor

☐ Sent to casualty   ☐ Admitted to hospital

Account of accident, dangerous occurrence or near miss
(Continued on separate sheet if necessary)

```
[                                                                    ]
[                                                                    ]
[                                                                    ]
[                                                                    ]
```

Witnesses to the incident
(Names, addresses and occupations)

```
[                                                                    ]
[                                                                    ]
[                                                                    ]
[                                                                    ]
```

Was the injured person wearing PPE? If yes, what PPE? _____

_____

Signature of person completing form _____

Occupation _____   **Date** _____

Figure 2.15 Accident/ near miss report

Activity

Think of a simple task you are familiar with and write a method statement for that task.

Method statement

Sometimes known as a 'safe system of work', a method statement details the way a task or process will be carried out safely. It includes a step-by-step guide, outlines the hazards involved, and describes the control measures that have been introduced to ensure the safety of anyone affected. Written method statements are often requested at the tender stage, so that the client can be sure of the company's safety credentials.

FAQ

How do I know what scale the drawing is at?

The scale should be written on the title panel (the box included on a plan or drawing giving basic information such as who drew it, how to contact them, the date and the scale).

How do I know if I need a schedule?

Schedules are only really used in large jobs where there is a lot of repeated design information. If your job has a lot of doors, windows, etc., it is a good idea to use one.

How do I know if I need approval?

If you are unsure, check section three of the Building Regulations or contact your local authority.

Do I need to know all the different Building Regulations and what is contained in each section?

No, but a good understanding of what is involved is needed.

How many different forms are there?

A lot of forms are used and some companies use more than others. You should ensure you get the relevant training on completing the form before using it.

Knowledge refresher

1   What is the difference between a timesheet and day worksheet?

2   What is a variation order?

3   What is a daily site diary used for?

4   Why are near miss reports used?

5   What is a method statement?

What would you do?

1   You are working on a renovation project when your boss calls you to ask what materials you need for the next few weeks. You are caught a bit off-guard, and you rush around giving your boss a list of materials over the phone. When the materials are delivered, there are some discrepancies: it's not what you said, as far as you can remember. You phone your boss to tell him and he gets cross, blaming you for the mistakes. Who is to blame? What should have been done?

2   A friend has approached you to do a loft conversion. You apply for planning permission and Building Regulations approval and are given both, so you start work. You come across a problem with the chimney and decide to remove some of the bricks. With the work completed, the Building Inspector shows up to check the job. What can the Building Inspector do? What effect could this have on the job? What could have been done to prevent it?

chapter 3

Planning and work programmes

OVERVIEW

Any building project begins long before the first brick is laid or the first foundation dug. Most buildings and construction projects will need some sort of planning approval before they get underway, as a range of planning restrictions are in place to keep building standards up, protect local people and protect the environment.

Work planning is also of paramount importance for every job, whether a single dwelling or a large housing estate. Without it even the smallest job can go wrong: something simple is forgotten or omitted, such as ordering a skip, and the job is suddenly delayed by anything up to a week. On a smaller job, poor planning can result in delays, which will harm your reputation and jeopardise future contracts. With larger contracts, penalty clauses can be costly: if the job overruns and isn't finished on time, the client may claim substantial amounts of money from the contractor.

This chapter will deal with:

- planning permission
- work programming.

Planning permission

Before starting to plan a building project, it is important to know how your plans may be affected by local and national building restrictions. The two main sets of restrictions you will come across are:

- the Building Regulations
- planning permission.

We looked at the Building Regulations in Chapter 2, so here we will look at planning permission. It is crucial that anyone planning a construction project understands how this works, and seeks the necessary approval in the correct way. If not, building work runs the risk of having to be halted, altered or even taken down.

Planning permission laws were introduced to stop people building whatever they like wherever they like. The submission of a planning application gives both the local authority and the general public a chance to look at the development, to see if it is in keeping with the local area and whether it serves the interests of the local community.

The main **remit** of planning laws is to control the use and development of land in order to obtain the greatest possible environmental advantages with the least inconvenience for both the person/s applying for permission and society as a whole.

The key word in planning is 'development', defined in planning law as 'the carrying out of building, engineering, mining or other operations in, on, over or under land, or the making of any material change in the use of any buildings or other land'. As well as building work, this covers the construction of a new road or driveway, and even change of use: if a bank is to be turned into a wine bar, planning permission will be needed.

Planning permission is required for most forms of development. Here are a few more examples of work requiring planning permission:

- virtually all new building work
- house extensions including conservatories, loft conversions and roof additions (such as dormers)
- buildings and other structures on the land including garages
- adding a porch to your house
- putting up a TV satellite dish.

Even if you are intending to work from home and wish to convert part of your home into an office, you will require planning permission if:

- your home is no longer to be used mainly as a private residence
- your business creates more traffic or creates problems with parking due to people calling

Definition

Remit – scope, job, the areas an organisation or individual has to cover

Did you know?

Planning permission is needed if you want to put up a satellite dish. The job itself is small and not disruptive, but a dish is thought to change the outer appearance of a house enough to need permission.

- your business involves any activities classed as unusual in a **residential** area

- your business disturbs your neighbours at unreasonable hours or creates other forms of nuisance or smell.

Not all work requires planning. You can make certain types of minor alterations to your house, such as putting up a fence or dividing wall (providing it is less than 1 metre high next to a highway, or under 2 metres elsewhere), without planning permission.

In areas such as conservation areas or classified Areas of Outstanding Natural Beauty there will be stricter controls on what is allowed. Listed buildings also have stricter controls and come under the Planning (Listed Buildings and Conservation Areas) Act 1990.

For planning permission, you must apply to your local council. When they look at your proposed works, they will take into consideration:

- the number, size, positioning, layout and external appearance of the buildings

- the proposed means of access, landscaping and impact on the neighbourhood

- **sustainability**, and whether the necessary infrastructure, such as roads, services, etc., will be available

- the proposed use of the development.

Several steps are involved in applying for planning permission. The first is to contact the local authority to see if they think planning permission is required (some councils may charge a small fee for this advice). If they say you do need planning permission, you need to then ask them for an application form. There are two types of planning permission that you can apply for:

- **Outline application** This can be made if you want to see what the council thinks of the building work you intend to do before you go to the trouble of having costly plans drawn up. Details of the work will have to be submitted later if the outline application is successful.

Definition

Residential – where people live, rather than a business district, for example

Definition

Sustainability– the ability to last or carry on, how easy something is to keep going

The public has a right to know about proposed developments

- **Full application** Here a full application is made with all the plans, specifications, and so on.

Once you have completed the relevant form this must be sent to the local authority along with any fee.

Next, the contents of your application will be publicised so that people can express their views and raise any objections. A copy will be placed in the planning register; an electronic version will be placed on the local authority's website; and immediate neighbours will be written to (or a fixed notice will be displayed on or as near as possible to the site). The local authority may also advertise your application in a local newspaper. As the applicant, you will be entitled to have a copy of any reports, objections and expressions of support the local authority receives regarding your application.

The local authority normally takes up to eight weeks to make a decision on your application but in some cases it may take longer. If this happens, the local authority should write asking for your written consent to extend the period. If your application is not dealt with within eight weeks, you can appeal to the Secretary of State, but this can be a lengthy procedure itself, so it is best to try to resolve the matter at a local level.

In looking at an application, the local authority considers whether there are valid reasons for refusing or granting permission: the local authority cannot simply reject a proposal because many people oppose it. The local authority will look at whether your proposal is consistent with the area's appearance, whether it will cause traffic problems and whether it has any impact on local amenities, environment and services.

Once an application has been looked at, there are four possible outcomes: permission refused; application still pending; granted with conditions; or granted.

- **Permission refused**

  If permission is refused, the local authority must state its reasons for turning down the application. If you feel these are unfair, you can appeal to the Secretary of State. Appeals must be made within six months of the local authority's decision and are intended as a last resort. It can take months to get a decision, which may be a refusal. Alternatively, you can ask what changes need to be made to allow the proposal to pass: if these are acceptable, the amended application can be submitted for processing. If after this the application is still rejected, the work cannot go ahead. However, different authorities have different procedures, so always check before submitting proposals.

- **Application still pending**

  Here the local authority may have found that it needs extra time to allow comments to come in, or to deal with particular issues that have arisen. If the application is still pending then, as stated previously, the local authority must ask for your written consent to extend the period for making a decision.

- **Granted with conditions**

In this case you are able to start the work, remembering to comply with the conditions stated. If you fail to comply, permission will be revoked and you may be ordered to undo the work done. If you are unhappy with the conditions set, you can ask for advice and, if needs be, make alterations to the plans. This would mean resubmitting the application.

- **Granted**

If you have been granted permission, you are free to start the work.

Knowledge refresher

1   Why were planning permission laws introduced?

2   What is the main remit of planning laws?

3   Give five examples of work that would require planning permission.

4   Give one example of work that would not require planning permission.

5   Give a brief outline of the two types of planning permission you could apply for.

6   List the four possible outcomes of a planning application.

What would you do?

You are working on a small job converting an attached garage into an extra bedroom. You have applied for planning permission and the application is still pending. The local authority say it should be fine, but there is one thing they need to check and it could take a few more weeks. This causes a problem for both you and the client: the client wants the work started and you have no other work to do for a few weeks. What do you do? What could the repercussions be? What could you do to protect yourself?

Work programming

Once planning permission and Building Regulations approval have been obtained, the next step is to plan the work (NB in some instances the client may ask the contractor to provide a work programme at the tender stage, to check the contractor's efficiency and organising ability).

A work programme is vital for good work planning, as it shows:

- what tasks are to be done and when, including any overlap in the tasks
- what materials are required and when
- what plant is needed, when and for how long
- what type of workforce is required and when.

A few different types of work programme are in use, and we will cover the main two here.

Planning the site

For every fair-sized job, the building site needs to be carefully planned. A poorly planned site can cause problems and delays, as well as incurring costs and even causing accidents.

A building site should be seen as a temporary workshop, store and office for the contractor, and must contain all the **amenities** needed on a permanent base. Sites should be planned in a way that minimises the movement of employees, materials and plant throughout the construction, while at the same time providing protection and security for employees, materials and components, and members of the public. A well-planned site will also have good transport routes, which will not disrupt the site or the general traffic.

Many things need to be included on a building site, so it is often easiest to plan your site using a site plan and cut-outs of the amenities you need. These cut-outs can be laid onto the plan and moved around until a suitable layout is found.

The ideal layout of the site will vary according to the size and **duration** of the job – there is no point hiring site offices for a job that will only last a day! The following gives an idea of what might be needed on an average site:

- **Site offices**

  The office space (usually portable cabins) should be of a decent size, usually with more than one room for different members of staff and a large room for meetings. Phone, fax and email facilities will be needed, so that the site office can communicate with Head Office, contractors, suppliers and others. As with any office, the site office must be heated, have plenty of light (natural or artificial) and be fitted out with useful, comfortable furniture.

Remember

If you need to plan several sites, save the cut-outs from one to use on the next (checking that you are using the same scale). You could end up with a 'kit' to use whenever you need it.

Definition

Amenities – facilities such as toilets, rest areas, etc.

Duration – how long something goes on

- **First aid office**

  This is sometimes contained within the site office, but on larger sites a separate space may be needed so that injured people can be treated quickly and efficiently. The first aid office must be fully stocked, and there must be sufficient trained first aiders on site.

- **Toilets**

  There must be sufficient toilets on the site. Usually there will be a WC block next to the canteen or mess area, with additional portable toilets dotted around the site if needed. Toilets must be kept clean and well stocked at all times, and have somewhere for people to wash their hands. The WC block may also need to house showers if the work being done requires them.

- **Lunch area**

  This should be protected from the wind and rain and have heating and electricity. It should contain equipment such as a microwave, kettle or urn and fridge to heat and keep food, as well as suitable food storage such as cupboards. There should be adequate seating and tables, and the space should be kept clean to prevent any unwelcome pests such as rats or cockroaches.

- **Drying room**

  This provides space for employees to dry off any clothes that get wet, on the way to or during work. It is usually sited next to the lunch area, or is part of the same building. The room must have adequate heating and ventilation, as well as lockers or storage to house things like motorcycle helmets.

- **Cranes, hoists, etc.**

  These can be static or portable. When a large static crane is required, its position needs to be planned so that it can easily and safely reach the area where it is needed. Larger cranes should be situated away from the main site office for safety reasons.

- **Transport route**

  Having a good transport route into, out of and within a building site is vital. It is best to have separate entrances and exits, with a one-way system on the site and good signposting throughout. These measures will avoid large delivery lorries having to turn around on site, and help to keep both internal and external traffic flowing with minimum disruption.

- **Waste area**

  This must be well away from the lunch area for health and safety reasons, and should be easily accessible from the transport route so that the skips and bins can be emptied easily. Separate well-labelled skips are needed for different kinds of refuse, and there should be some for recycling. Certain skips should be kept separate to avoid **contamination**, and chemical dumps (for paint, etc.) should be kept secure and emptied regularly.

Various types of storage are also needed on a building site, such as:

- **materials storage** – enough adequate space to store all types of materials, ideally near to where they are being used (for example, cement and sand should be stored near the mixer). All materials should be stored in a way that prevents them being damaged or stolen; some materials will have to be stored separately to avoid contamination.

- **component storage** – a secure compound protected from the wind and rain for items such as doors and windows. Again, components should be stored in a way that prevents them being damaged.

- **tool storage** – a secure place for employees' own tools as well as site tools such as table saws. The tool storage area needs to be thoroughly secure to prevent theft.

- **ironmongery storage** – a locked compound in a container with well-labelled racks to avoid things like screws and nails being mixed up. Expensive ironmongery such as door furniture needs to be properly secure. On a well-planned site, expensive ironmongery is only ordered when needed.

A good site layout might look something like this.

Activity

Design a site layout for a large project, placing all the amenities that you think may be required in an appropriate place.

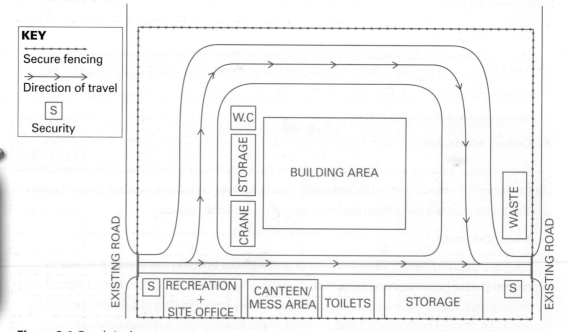

Figure 3.1 Good site layout

Planning the work

There are many types of work programme, including the critical path and the Bar/Gantt chart. The latter is the one you will come across most often.

Bar charts

The bar or Gantt chart is the most popular work programme as it is simple to construct and easy to understand. Bar charts have tasks listed in a vertical column on the left and a horizontal timescale running along the top.

Time in days										
Activity	1	2	3	4	5	6	7	8	9	10
Dig for foundation and service routes										
Lay foundations										
Run cabling, piping, etc. to meet existing services										
Build up to DPC										
Lay concrete floor										

Figure 3.2 Basic bar chart

Each task is given a proposed time, which is shaded in along the horizontal timescale. Timescales often overlap as one task often overlaps another.

Time in days										
Activity	1	2	3	4	5	6	7	8	9	10
Dig for foundation and service routes	██	██								
Lay foundations			██	██						
Run cabling, piping, etc. to meet existing services				██	██					
Build up to DPC						██	██			
Lay concrete floor								██	██	██

Key: proposed ██

Figure 3.3 Bar chart showing proposed time for a contract

The bar chart can then be used to check progress. Often the actual time taken for a task is shaded in underneath the proposed time (in a different way or colour to avoid confusion). This shows how what *has* been done matches up to what *should have* been done.

Did you know?

The Gantt chart is named after the first man to publish it. This was Henry Gantt, an American engineer, in 1910.

Activity

Think of a task and create a bar chart for that task.

58

Activity

Using the bar chart you've created, think of what could go wrong and write a list of contingencies to overcome any problems.

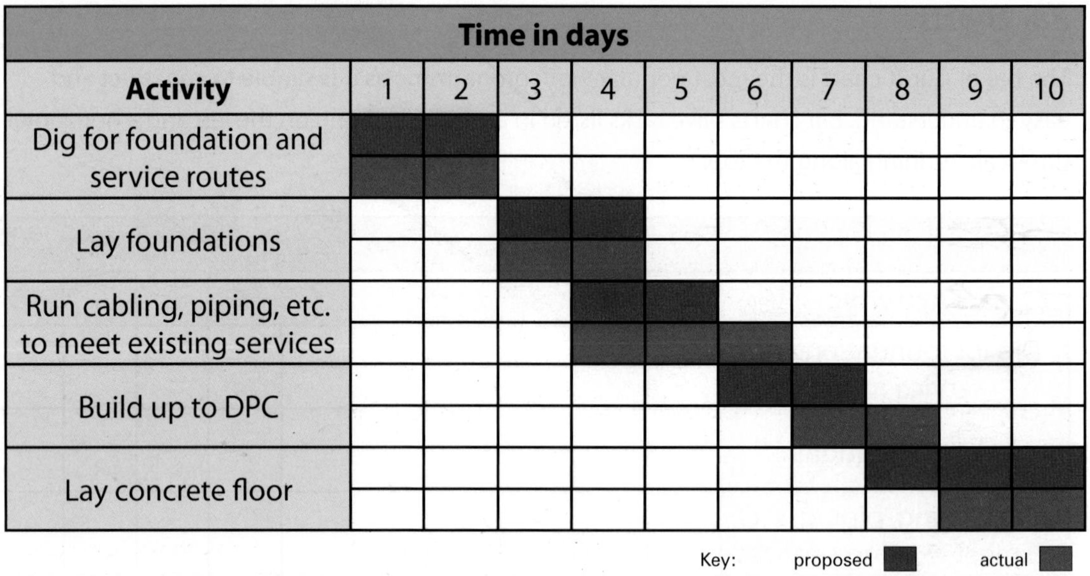

Time in days										
Activity	1	2	3	4	5	6	7	8	9	10
Dig for foundation and service routes	■	■								
	■	■								
Lay foundations			■	■						
			■	■						
Run cabling, piping, etc. to meet existing services				■	■					
				■	■					
Build up to DPC						■	■			
							■	■		
Lay concrete floor								■	■	■
									■	■

Key: proposed ■   actual ■

Figure 3.4 Bar chart showing actual time half way through a contract

As you can see, a bar chart can help you plan when to order materials or plant, see what trade is due in and when, and so on. A bar chart can also tell you if you are behind on a job; if you have a penalty clause written into your contract, this information is vital.

When creating a bar chart, you should build in some extra time to allow for things such as bad weather, labour shortages, delivery problems or illness. It is also advisable to have contingency plans to help solve or avoid problems, such as:

- capacity to work overtime to catch up time

- bonus scheme to increase productivity

- penalty clause on suppliers to try to avoid late or poor deliveries

- source of extra labour (e.g. from another site) if needed.

Good planning, with contingency plans in place, should allow a job to run smoothly and finish on time, leading to the contractor making a profit.

Critical paths

Another form of work programme is the critical path. Critical paths are rarely used these days as they can be difficult to decipher. The final part of this chapter will give a brief overview of the basics of a critical path, in case you should come across one.

A critical path can be used in the same way as a bar chart to show what needs to be done and in what sequence. It also shows a timescale but in a different way to a bar chart: each timescale shows both the minimum and the maximum amount of time a task might take.

The critical path is shown as a series of circles called event nodes. Each node is split into three: the top third shows the event number, the bottom left shows the earliest start time, and the bottom right the latest start time.

Did you know?

Bad weather is the main external factor responsible for delays on building sites in the UK. A Met Office survey showed that the average UK construction company experiences problems caused by the weather 26 times a year.

The nodes are joined together by lines, which represent the tasks being carried out between those nodes. The length of each task is shown by the times written in the lower parts of the nodes. Some critical paths have information on each task written underneath the lines that join the nodes, making them easier to read.

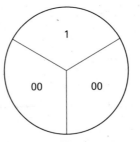

Figure 3.5 Single event node

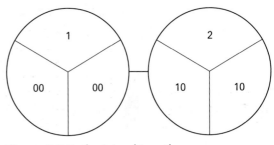

Figure 3.6 Nodes joined together

On a job, many tasks can be worked on at the same time, e.g. the electricians may be wiring at the same time as the plumber is putting in his pipes. To show this on a critical path, the path can be split.

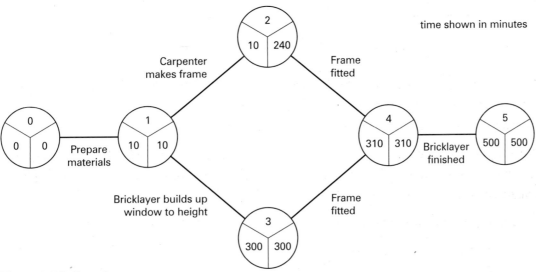

time shown in minutes

Figure 3.7 Split path

The example shown shows how a critical path can be used for planning building in a window opening, with a carpenter creating a dummy frame.

The event nodes work as follows:

- **Node 0** – This is the starting point.

- **Node 1** – This is the first task, where the materials are prepared.

- **Node 2** – This is where the carpenter makes the dummy frame for the opening. Notice that the earliest start time is 10 minutes and the last start time is 240 minutes. This means that the carpenter can start building the frame at any time between 10 minutes and 240 minutes into the project. This is because the frame will not be needed until 300 minutes, but the job will only take 60 minutes. If the carpenter starts *after* 240 minutes, there is a possibility that the job may run behind.

- **Node 3** – This is where the bricklayer must be at the site, ready for the frame to be fitted at 300 minutes, or the job will run behind.

Activity

Think of a task you are familiar with and create a critical path for that task.

- **Node 4** – With the frame fitted, the bricklayer starts at 310 minutes and has until node 5 (500 minutes) to finish.

- **Node 5** – The job should be completed.

When working with a split path it is vital to remember that certain tasks have to be completed before others can begin. If this is not taken into account on the critical path, the job will run over (which may prove costly, both through penalty clauses and also in terms of the contractor's reputation).

On a large job, it can be easy to misread a critical path as there may be several splits, which could lead to confusion.

Remember

Whichever way you choose to programme your work, your programme must be realistic, with clear objectives and achievable goals.

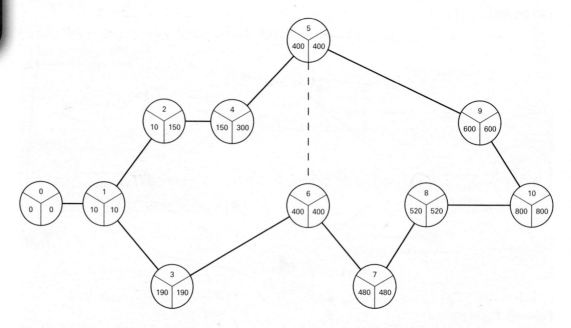

Figure 3.8 Critical path for a large job

FAQ

How do I know if my job needs planning permission?

If you are unsure, you should contact your local council.

What type of planning permission should I apply for?

If you are unsure of your work, you can make an outline application, which will tell you if your job will pass without getting costly plans made up (though you will have to submit plans later). If you are confident of what you want, you can apply for a full application.

How much does planning permission cost?

The costs vary depending on what application you make and to which council you make it.

Do I need to have all the listed amenities on my building site?

No. The amenities listed are a guide to what should be on a large site. If you are just doing an extension, the amenities needed will be fewer and simpler (e.g. no site office).

Which type of programme should I use: bar chart or critical path?

It is up to the individual which programme they use – both have their good points – but a bar chart is the easiest to set up and work from.

What if it rains for the entire 20-day duration of the job?

The job would be seriously behind schedule. You can't plan for the weather in this country, but it would be unwise to start this job during a rainy season. There are companies that can provide scaffolding with a fitted canopy to protect the work area, which would be ideal for a job of this size. Larger jobs have longer programmes, and when they are drawn up they are made more flexible to allow for a lot of rainy days.

Knowledge refresher

1   List four things that might be included in the layout for a large site.

2   Why is it best to have a good transport route within a site?

3   List four different types of storage that may be needed on a building site.

4   State four pieces of information you can get from a bar chart.

5   With regard to critical paths, what three things are contained in an event node?

What would you do?

You have been tasked with designing a programme of work for a large contract involving the building of 20 houses. What sort of thing should you check prior to starting? What should you do about plant, labour, materials? What sort of programme should you use (bar or critical path)? What amenities should you consider?

chapter 4

Drawings

OVERVIEW

When drawings are mentioned in the construction industry, people generally tend to think of the architect's drawings and plans that form part of the contract documents. These types of drawings are vital in the construction industry as they form part of the legal contract between client and contractor – and mistakes, either in design or interpretation of the design, can be costly.

However, there are other forms of drawings that are just as important. Setting-out drawings are used to mark out for complex procedures such as constructing cut roofing, staircases or brick arches; and with advances in technology, CAD (computer-aided design) is being used more often.

The Level 2 book gave a good grounding in contract document drawings. This chapter will give you a refresher, and will then expand on your knowledge in this area as well as looking at the wider range of drawings involved in construction today.

This chapter will cover:

- types of drawing
- setting out drawings
- projections
- computer-aided design.

Types of drawing

Plans and drawings are vital to any building work as a way of expressing the client's wishes. Drawings are the best way of communicating a lot of detailed information without the need for pages and pages of text. Drawings form part of the contract documents and go through several stages before they are given to tradespeople for use.

Stage 1 The client sits down with an architect and explains his/her requirements.

Stage 2 The architect produces drawings of the work and checks with the client to see if the drawings match what the client wants.

Stage 3 If required, the drawings go to planning to see if they can be allowed, and are also scrutinised by the Building Regulations Authority. It is at this stage that the drawings may need to be altered to meet Planning or Building Regulations.

Stage 4 Once passed, the drawings are given to contractors along with the other contract documents, so that they can prepare their tenders for the contract.

Stage 5 The winning contractor uses the drawings to carry out the job. At this point the drawings will be given to you to work from.

There are three main types of working drawings: location drawings, component drawings and assembly drawings. We will look at each of these in turn.

Location drawings

Location drawings include:

- **block plans**, which identify the proposed site in relation to the surrounding area. These are usually drawn at a scale of 1:2500 or 1:1250

Activity

Produce a block plan containing the building you are in and the surrounding buildings roads, etc.

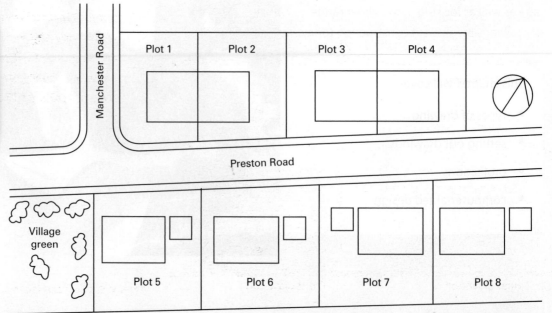

Figure 4.1 Block plan

- **site plans**, which give the position of the proposed building and the general layout of things such as services and drainage. These are usually drawn at a scale of 1:500 or 1:200

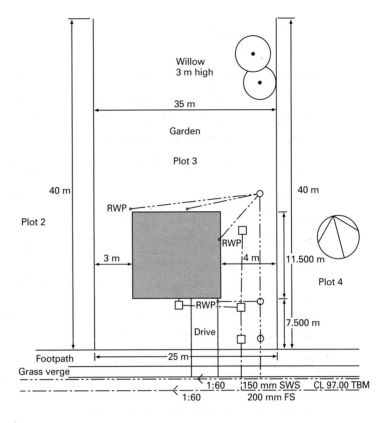

Figure 4.2 Site plan

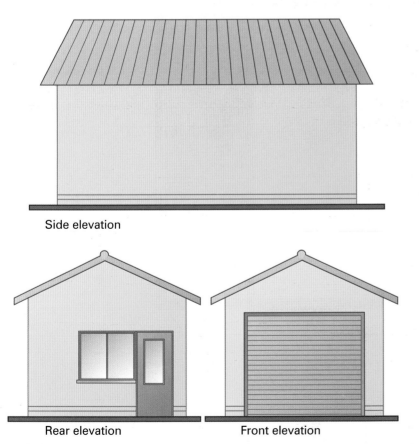

Side elevation

Rear elevation

Front elevation

- **general location drawings**, which show different elevations and sections of the building. These are usually drawn at a scale of 1:200, 1:100 or 1:50

Figure 4.3 General location drawing

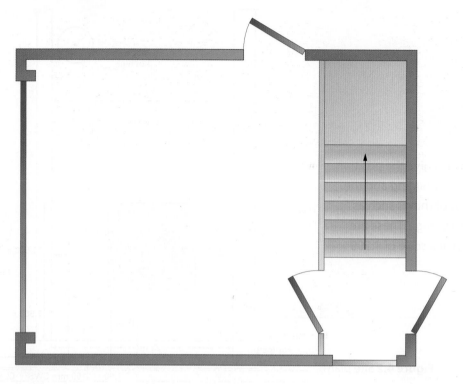

Ground floor plan

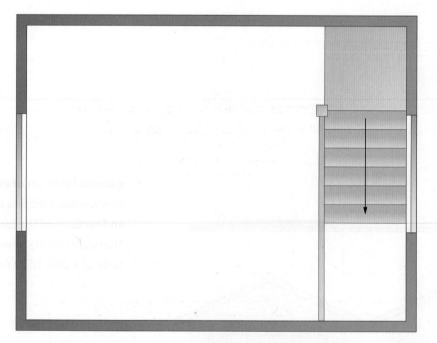

First floor plan

Figure 4.4 Floor plans

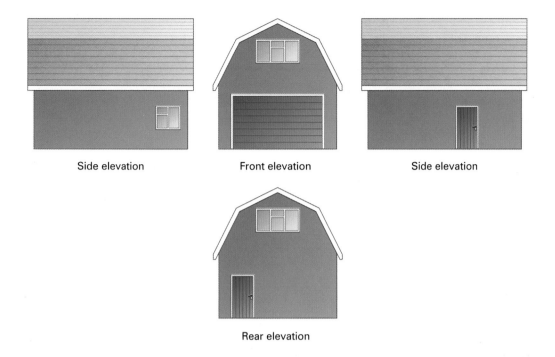

Figure 4.5 Elevation drawings

Component drawings

Component drawings include:

- **range drawings**, which show the different sizes and shapes of a particular range of components. These are usually drawn at a scale of 1:50 or 1:20

Activity

Produce a component drawing for an item that you are familiar with.

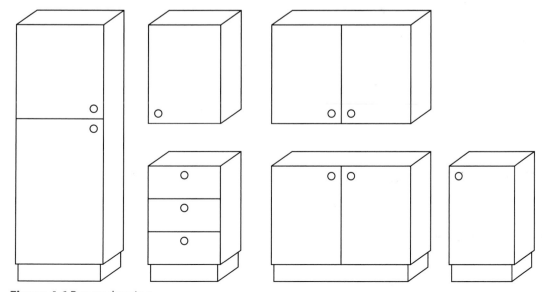

Figure 4.6 Range drawing

- **detailed drawings**, which show all the information needed to complete or manufacture a component. These are usually drawn at a scale of 1:10, 1:5 or 1:1.

Activity

Produce a detailed drawing of a component you are familiar with.

Drip mould

Head

Top rail

Bottom rail

Drip mould

Transom

Mortar key

Sub sill

Anti-capillary groove

Sill

Window board

Figure 4.7 Detailed drawing

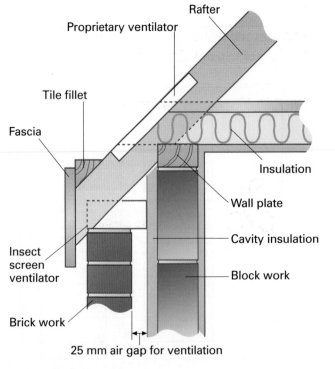

Rafter

Proprietary ventilator

Tile fillet

Fascia

Insulation

Wall plate

Cavity insulation

Insect screen ventilator

Block work

Brick work

25 mm air gap for ventilation

Figure 4.8 Assembly drawing

Assembly drawings

Assembly drawings are similar to detailed drawings and show in great detail the various joints and junctions in and between the various parts and components of a building. Assembly drawings are usually drawn at a scale of 1:20, 1:10 or 1:5.

All plans and drawings contain symbols and abbreviations, which are used to show the maximum amount of information in a clear and legible way.

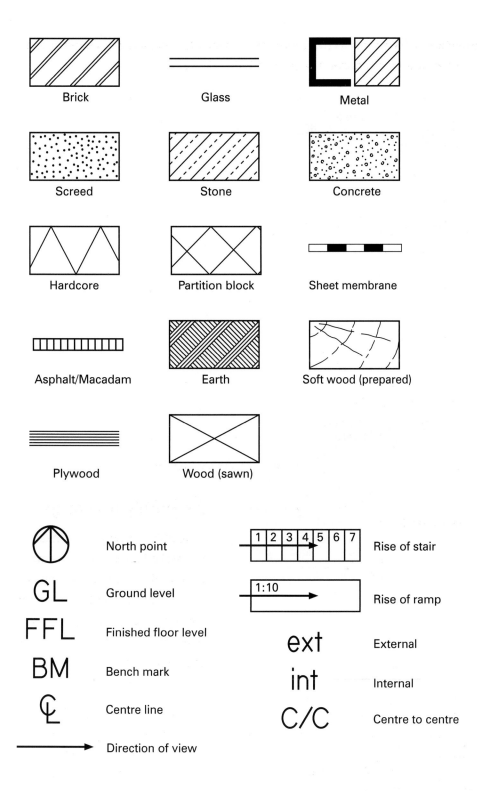

Figure 4.9 Symbols

Item	Abbreviation	Item	Abbreviation
Airbrick	AB	Hardcore	hc
Asbestos	abs	Hardwood	hwd
Bitumen	bit	Insulation	insul
Boarding	bdg	Joist	jst
Brickwork	bwk	Mild steel	MS
Building	bldg	Plasterboard	pbd
Cast iron	CI	Polyvinyl acetate	PVA
Cement	ct	Polyvinyl chloride	PVC
Column	col	Reinforced concrete	RC
Concrete	conc	Satin chrome	SC
Cupboard	cpd	Satin anodised aluminium	SAA
Damp proof course	DPC	Softwood	swd
Damp proof membrane	DPM	Stainless steel	SS
Drawing	dwg	Tongue and groove	T&G
Foundation	fnd	Wrought iron	WI
Hardboard	hdbd		

Table 4.1 Abbreviations

Knowledge refresher

1 State what happens at the third stage of the client/architect consultation.

2 What is a suitable scale for a block plan?

3 Give a suitable scale for a general location drawing.

4 Describe what a component range drawing is.

What would you do?

1 You are working on a job and have received the site plans, which show the layout of the services. You start to dig out for the services and, when you reach the site where the mains gas should be, you find it is not where the drawing shows. What could have caused this problem? What further problems could be caused? What effect could this have financially?

2 You have been issued a scale drawing for building internal walls, but some of the dimensions are missing. What should you do? What complications could arise from scaling from the drawing as it is? What effect could building a wall in the wrong place have?

Setting out drawings

Settings out drawings are as important as contract documents. You must be aware of how certain tasks are set out and what drawings can be created to aid in the setting out process.

The setting out drawings are most often needed on smaller jobs, where there is limited or no information from the architect in the form of contract document drawings. Setting out drawings can also be used on larger sites where there has been an alteration or on oversight by the architect.

Here is where the most common forms of setting out drawings are used:

- in carpentry, for cut roofing, where there may be no information on the true lengths of rafters

- in joinery, when setting out for stairs, where there may be no information on the individual rise, etc.

- in bricklaying, where you may come across setting out drawing for arch centres, such as segmental or gothic arches.

Setting out drawings are crucial for creating arches like these

We will now look at a brief example of how roofing and brick arches are set out.

Finding the true length of a common rafter

Most drawings will tell you the **span** and **rise** of the roof. From these, you can create a drawing that will tell you the true length of the common rafter, and also what angle the ends of the rafter should be cut at.

This true length is the actual length that the rafter needs to be, and all the rafters can be cut to length from the setting out drawing. The setting out drawing for a roof is usually drawn on a sheet of plywood to a scale that fits the sheet.

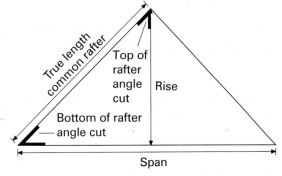

Figure 4.10 Finding common rafter true length

Definition

Span – the distance measured in the direction of ceiling joists, from the outside of one wall plate to another, known as the overall (O/A) span.

Rise – the distance from the top of the wall plate to the roof's peak

Setting out a segmental brick arch

Most drawings will show you the opening span of the arch, but some may not tell you the radius. Without the radius, you cannot build the arch correctly.

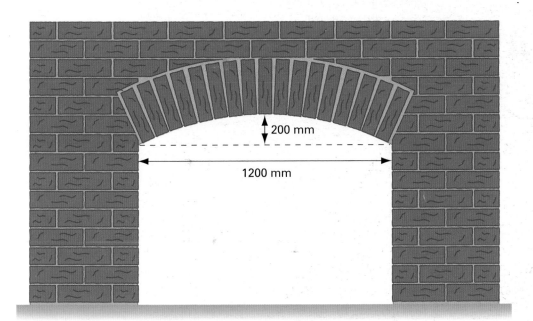

Figure 4.11 An example of a segmental arch

Activity

Using a suitable scale, create a setting out drawing for a rafter with a span of 3.5 m and a rise of 1.4 m, so that the true length of the rafter and angles of cuts can be identified.

We will now look at how setting out drawings can aid you in setting out this arch.

Figure 4.12 Establish the span (a length of 1200 mm has been used here, shown as A–B)

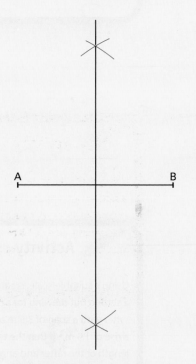

Figure 4.13 Bisect this line

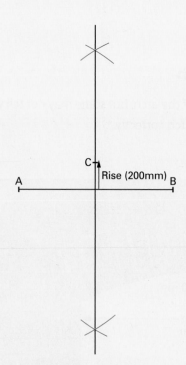

Figure 4.14 Establish the rise (the distance from the springing line (A–B) to the highest point of the soffit shown as C). The rise is normally one sixth of the span, so in this case, the rise is shown as 200 mm

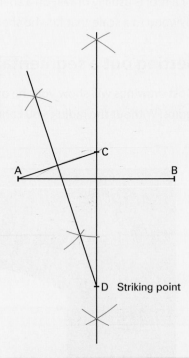

Figure 4.15 Draw a line from A to C and bisect this line. The point where this bisecting line crosses the bisecting line of the span will be the striking point for the arch (shown here as point D)

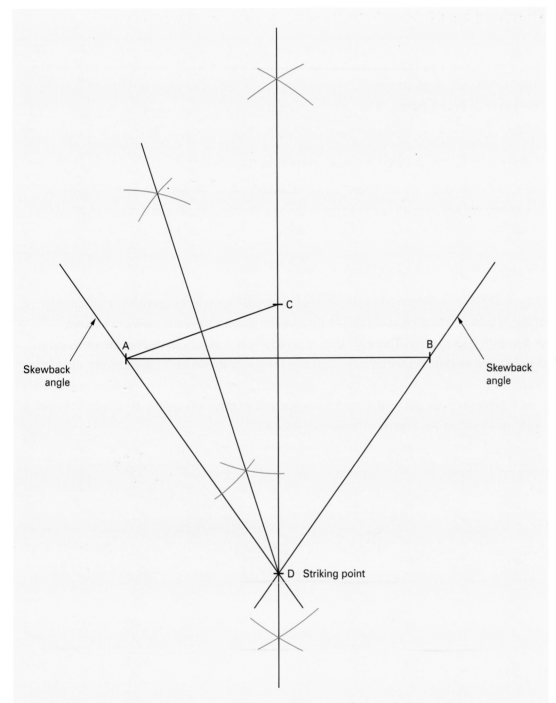

Figure 4.16 Draw a line from D through A and a line from D through B. These lines will provide the angle for the **skewbacks**.

Definition

Skewbacks – the angle at the springing point at which the arch rings will be laid

Setting out for a segmental arch can again be drawn out on a sheet of plywood but, in this case, it can be drawn full size, with the drawing being cut out and used as a template for the arch centre.

Projections

Building, engineering and similar drawings aim to give as much information as possible in a way that is easy to understand. They frequently combine several views on a single drawing. These may be of two kinds:

- elevation – the view we would see if we stood in front or to the side of the finished building

- plan – the view we would have if we were looking down on it.

The view we see depends on where we are looking from. There are then different ways of 'projecting' what we would see onto the drawings. The three main methods of projection, used on standard building drawings, are orthographic, isometric and oblique.

Orthographic projection

Orthographic projection works as if parallel lines were drawn from every point on a model of the building on to a sheet of paper held up behind it (an elevation view), or laid out underneath it (plan view). There are then different ways that we can display the views on a drawing. The method most commonly used in the building industry, for detailed construction drawings, is called 'third angle projection'. In this the front elevation is roughly central. The plan view is drawn directly below the front elevation and all other elevations are drawn in line with the front elevation. An example is shown in Figure 4.17.

Front elevation

Side elevation

Figure 4.17 Orthographic projection

Isometric projection

In isometric views, the object is drawn at an angle where one corner of the object is closest to the viewer. Vertical lines remain vertical but horizontal lines are drawn at an angle of 30° to the horizontal. This can be seen in Figure 4.18.

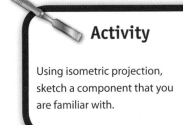

Activity

Using isometric projection, sketch a component that you are familiar with.

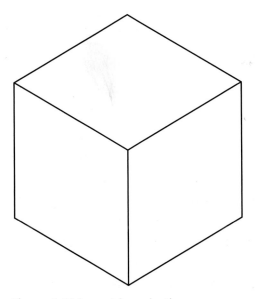

Figure 4.18 Isometric projection

Oblique projection

Oblique projection is similar to an isometric view, with the object drawn at an angle where one corner of the object is closest to the viewer. Vertical lines remain vertical but horizontal lines are drawn at an angle of 45° to the horizontal. This can be seen in Figure 4.19.

Activity

Using oblique projection, sketch a component that you are familiar with.

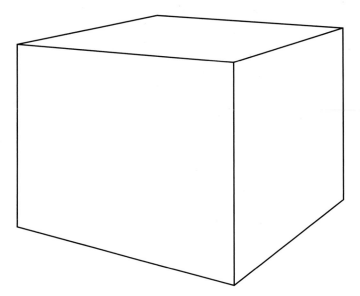

Figure 4.19 Oblique projection

Knowledge refresher

1   What is the purpose of setting out drawings?

2   What are the main types of projection used in construction drawings?

3   In isometric projection, at what angle are the horizontal lines drawn?

What would you do?

You have been tasked with building a segmental brick arch, but there is minimal information on the drawing. You decide to just build the arch but soon run into problems with the radius. What could have prevented the problems? What should you do now? What effect can this have on the building and profitability of the job?

Computer-aided design

Computer-aided design (CAD) is a system in which a draftsperson uses computer technology to help design a part, product or whole building. It is both a visual and symbol-based method of communication, with conventions particular to a specific technical field.

CAD is used particularly at the drafting stage. Drafting can be done in two dimensions (2-D) and three dimensions (3-D).

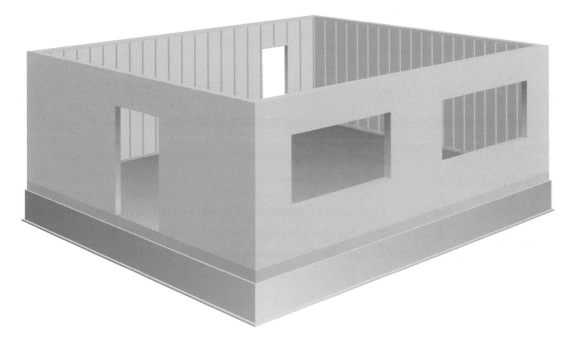

Figure 4.20 A CAD drawing

CAD is one of the many tools used by engineers and designers, and is used in many ways, depending on the profession of the user and the type of software in question.

There are several types of CAD. Each requires the operator to think differently about how he or she will use it, and he or she must design their virtual components in a different manner for each.

Many companies produce lower-end 2-D systems, and a number of free and open source programs are available. These make the drawing process easier, because there are no concerns about the scale and placement on the drawing sheet that accompanied hand drafting – these can simply be adjusted as required during the creation of the final draft.

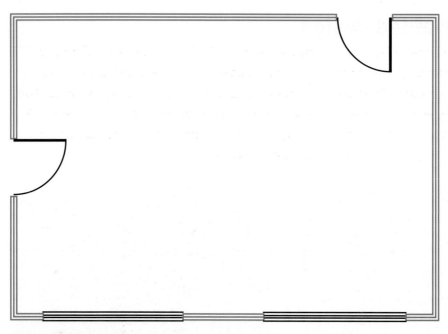

Figure 4.21 A 2-D CAD drawing

3-D wireframe

3D wireframe is in essence an extension of 2-D drafting. Each line has to be manually inserted into the drawing. The final product has no mass properties associated with it, and cannot have features directly added to it, such as holes. The operator approaches these in a similar fashion to the 2-D systems, although many 3-D systems allow you to use the wireframe model to make the final engineering drawing views.

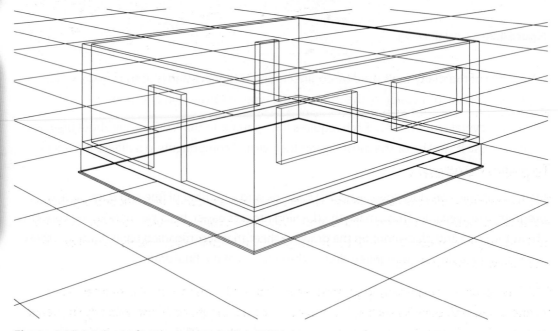

Figure 4.22 A 3-D wireframe produced using CAD

3-D dumb solids

3D 'dumb' solids are created in a way corresponding to manipulations of real-world objects. Basic three-dimensional geometric forms (prisms, cylinders, spheres, and so on) have solid volumes added to or subtracted from them, as if assembling or cutting real-world objects. Two-dimensional projected views can easily be generated from the models. The sorts of basic 3-D solids that are created do not usually include tools to easily allow motion of components, set limits to their motion, or identify interference between components.

Figure 4.23 A 3-D view of a house produced using CAD

Top-end systems

Top-end systems offer the capabilities to incorporate more organic, aesthetic and ergonomic features into designs. Freeform surface modelling is often combined with solids to allow the designer to create products that fit the human form and visual requirements, as well as the interface with the machine.

CAD has become an especially important technology within the scope of computer-aided technologies, with benefits such as lower product development costs and a greatly shortened design cycle. CAD enables designers to lay out and develop work on screen, print it out and save it for future editing, saving time on their drawings.

Knowledge refresher

1 What is CAD?

2 Describe a 3-D wireframe program.

3 Describe a 3-D dumb solids program.

4 What are the main advantages of using a CAD system?

chapter 5

Building methods and construction technologies

OVERVIEW

Whatever type of building you may be involved in constructing, there are certain elements that must be included and certain principles that must be followed. For example, a block of flats and a warehouse will both have foundations, a roof, and so on.

At Level 2, you learned about the basic elements of a building. In this chapter, you will look in greater depth at the main elements and principles of building work and the materials used.

This chapter should be read in conjunction with Chapter 6, which looks specifically at the energy efficiency and sustainability of different building methods and materials.

This chapter will cover:

- foundations
- exterior walls
- internal walls
- floors
- roofs.

Insulation, which is now an important aspect of walls, flooring and roofing, is dealt with in Chapter 6.

Foundations

Any building work will start with the foundations. The design of any foundation will depend on a number of factors including ground conditions, soil type, the location of drains and trees in relation to the building, and any loads that may be generated, either by the structure or naturally.

The purpose of foundations

A building damaged through subsidence

> ### Definition
>
> **Dead load** – the weight of the structure
>
> **Imposed load** – the additional weight/loading that may be placed on the structure itself

The foundations of a building ensure that all **dead** and **imposed loads** are safely absorbed and transmitted through to the natural foundation or sub-soils on which the building is constructed. Failure to adequately absorb and transmit these loads will result in the stability of the building being compromised, and will undoubtedly cause structural damage.

Foundations must also be able to allow for ground movement brought about by the soil shrinking as it dries out and expanding as it becomes wet. The severity of shrinkage or expansion depends on the type of soil you are building on.

Frost may also affect ground movement, particularly in soils that hold water for long periods. When this retained water freezes, it can make the sub-soil expand.

Tree roots and future excavations can also cause movement that affects the sub-soil.

Types of soil

As you can imagine there are many different types of soils. For foundation design purposes, these have been categorised as follows:

- rock
- gravel
- clay
- sand
- silt.

Each of these categories of sub-soil can be broken down even further: for example,

- clay which is sandy and very soft in its composition
- clay which is sandy but very stiff in its composition.

This information will be of most interest to the architect, but nonetheless is of the utmost importance when designing the foundation.

A number of calculations are used to determine the size and make-up of the foundation. These calculations take into account the **load-bearing capacity of the subsoil**. Calculations for some of the more common types of foundations can be found in the current *Building Regulations*. However, these published calculations cannot possibly cover all situations. Ultimately it will be down to the expertise of the building design teams to accurately calculate the bearing capacity of soils and the make-up of the foundation.

In the early stages of the design process, before any construction work begins, a site investigation will be carried out to ascertain any conditions, situations or surrounding sites which may affect the proposed construction work. A great deal of data will need to be established during site investigations, including:

- position of boundary fences and hedges
- location and depth of services, including gas, electricity, water, telephone cables, drains and sewers
- existing buildings which need to be demolished or protected
- position, height, girth and spread of trees
- types of soil and the depths of these various soils.

The local authorities will normally provide information relating to the location of services, existing buildings, planning restrictions, preservation orders and boundary demarcation. However, all of these will still need to be identified and confirmed through the site investigation. In particular, hidden services will need to be located with the use of modern electronic surveying equipment.

Definition

Load-bearing capacity of subsoil – the load that can be safely carried by the soil without any adverse settlement

Find out

Look at the different methods and equipment used to locate and identify various hidden services.

Find out

How are the different soil tests carried out?

Did you know?

Site investigations or surveys will also establish the contours of the site. This will identify where certain areas of the site will need to be reduced or increased in height. An area of the site may need to be built up in order to mask surrounding features outside the boundaries of the proposed building project.

Activity

In an area designated by your trainer, carry out a site survey and record your results.

Find out

How can plant growth affect some structures?

Soil investigations are critical. Samples of the soil are taken from various points around the site and tested for their composition and for any contamination. Some soils contain chemicals that can seriously damage the foundation concrete. These chemicals include sodium and magnesium sulphates. The effects of these chemicals on the concrete can be counteracted with the use of sulphate-resistant cements.

Many different tests can be carried out on soil. Some are carried out on site; others need to be carried out in laboratories. Tests on soil include:

- penetration tests – to establish density of soil
- compression tests – to establish shear strength of the soil or its load-bearing capacity
- various laboratory tests – to establish particle size, moisture content, humus content and chemical content.

Once all site investigations have been completed and all necessary information and data has been established in relation to the proposed building project, site clearance can take place.

Site clearance

The main purpose of site clearance is to remove existing buildings, waste, vegetation and, most importantly, the surface layer of soil referred to as topsoil. It is necessary to remove this layer of soil, as it is unsuitable to build on. This surface layer of soil is difficult to compact down due to the high content of vegetable matter, which makes the composition of the soil soft and loose. The topsoil also contains various chemicals that encourage plant growth, which may adversely affect some structures over time.

The process of removing the topsoil can be very costly, in terms of both labour and transportation. The site investigations will determine the volume of topsoil that needs to be removed.

In some instances, the excavated topsoil may not be transported off site. Where building projects include garden plots, the topsoil may just need to be stored on site, thus reducing excessive labour and transportation costs. However, where this is the case, the topsoil must be stored well away from areas where

Removing soil from a site

buildings are to be erected or materials are to be stored, to prevent contamination of soils or materials.

Once the site clearance is complete, excavations for the foundations can start.

Trench excavation

In most modern-day construction projects, trenches are excavated by mechanical means. Although this is an expensive method, it reduces labour time and the risks associated with manual excavation work. Even with the use of machines to carry out excavations, an element of manual labour will still be needed to clean up the excavation work: loose soil from both the base and sides of the trench will have to be removed, and the sides of the trench will have to be finished vertically.

Manual labour is still required for excavating trenches on some projects where machine access is limited and where only small strip foundations of minimum depths are required.

Trenches to be excavated are identified by lines attached to and stretched between profiles. This is the most accurate method of ensuring trenches are dug to the exact widths.

Excavation work must be carefully planned as workers are killed or seriously injured every year while working in and around trenches. Thorough risk assessments need to be carried out and method statements produced prior to any excavation work commencing.

Potential hazards are numerous and include: possible collapse of the sides of the trench, hitting hidden services, plant machinery falling into the trench and people falling into the trench.

One main cause of trench collapse is the poor placement of materials near to sides of the trench. Not only can materials cause trench collapse, but they may also fall into the trench onto workers. Materials should not be stored near to trenches. Where there is a need to place materials close to the trench for use in the trench itself, always ensure these are kept to a minimum, stacked correctly and used quickly and, most importantly, ensure the trench sides are supported.

Trenches are often excavated by mechanical means

Trench support

The type and extent of support required in an excavated trench will depend predominantly on the depth of the trench and the stability of the subsoil.

Traditionally, trench support was provided just by using varying lengths and sizes of timber, which can easily be cut to required lengths. However, timber can become unreliable under certain loadings, pressures and weather conditions and can fail in its purpose.

More modern types of materials have been introduced as less costly and time-consuming methods of providing the required support. These materials include steel sheeting, rails and props. Trench support can be provided with a mixture of both timber and steel components.

Here you can see the methods of providing support in trenches with differing materials and a combination of these materials.

Did you know?

The Health and Safety Executive has produced detailed documents that deal exclusively with safety in excavations. These documents can be downloaded from the HSE website or obtained upon request direct from the HSE.

Regulations relating to safety in excavations are set out in the *Construction Regulations* and these must be strictly adhered to during the work.

Activity

Using sketches, show the different ways of supporting trench excavations.

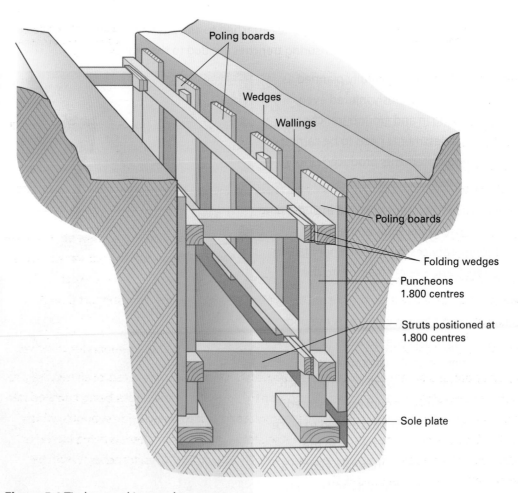

Figure 5.1 Timber used in trench support

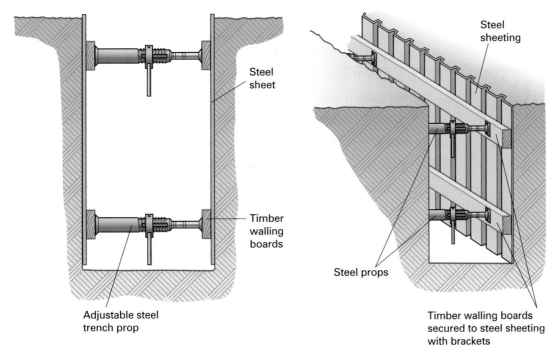

Figure 5.2 Combination of timber and steel used in trench support

The amount of timber or other materials required to provide adequate temporary support will be determined by the characteristics of the soil and the soil's ability to remain stable during the time over which the work is carried out. The atmospheric conditions will also affect the soil's ability to remain stable. The longer the soil is exposed to the natural elements, the more chance there is of the soil shrinking or expanding.

Without support, soil will have a natural angle of repose: in other words, the angle at which the soil will rest without collapsing or moving. Again, this will be affected by the natural elements to which the soil is exposed. It is virtually impossible to accurately establish the exact angle at which a type of soil will settle, so it is always advisable to provide more support than is actually required.

Site engineers will carry out calculations in relation to the support requirements for trenches.

Temporary barriers or fences should also be provided around the perimeters of all trenches, to prevent people falling into the trenches and also to prevent materials from being knocked into them. Good trench support methods will incorporate extended trench side supports, which provide a barrier – similar to a toe board on a scaffold – to prevent materials being kicked or knocked into the trench. Where barriers or fences are impractical, then trenches should be covered with suitable sheet materials.

In addition to the supports already mentioned, any services which run through the excavated trenches (in particular drains and gas pipes) need to be supported, especially where the ground has to be excavated underneath them.

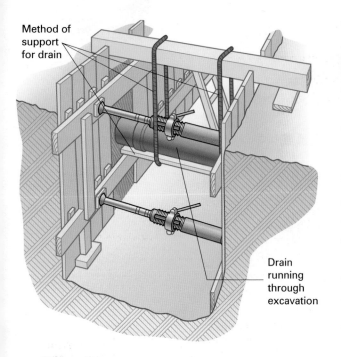

Method of support for drain

Drain running through excavation

Figure 5.3 Support for drains running through an excavation

Where trenches have to be excavated close to existing buildings, it may be necessary to provide support to the elevation adjacent to the excavation. This is due to the fact that, as ground is taken away from around the existing foundations, the loads will not be adequately and evenly distributed and absorbed into the natural or sub-foundation, possibly causing the structure to collapse. This support is known as shoring.

One other factor that can affect the safety of workers in excavations and the stability of the soil is surface water. Surface water can be found at varying levels within the soil and, depending on the depth, trenches can easily cause flooding. Where this occurs, water pumps will need to be used to keep the trench clear. Failure to keep the trench free of water during construction will not only make operations difficult, but may also weaken and loosen the support systems due to soil displacement.

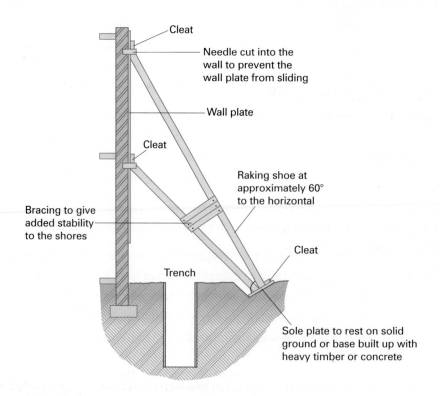

Cleat

Needle cut into the wall to prevent the wall plate from sliding

Wall plate

Cleat

Raking shoe at approximately 60° to the horizontal

Bracing to give added stability to the shores

Cleat

Trench

Sole plate to rest on solid ground or base built up with heavy timber or concrete

Figure 5.4 Raking shores used to support an existing building

Types of foundation

As previously stated, the design of a foundation will be down to the architect and structural design team. The final decision on the suitability and depth of the foundation, and on the thickness of the concrete, will rest with the local authority's building control department.

Strip foundations

The most commonly used strip foundation is the 'narrow strip' foundation, which is used for small domestic dwellings and low-rise structures. Once the trench has been excavated, it is filled with concrete to within 4–5 courses of the ground level DPC. The level of the concrete fill can be reduced in height, but this makes it difficult for the bricklayer due to the confined area in which to lay bricks or blocks.

The depth of this type of foundation must be such that the subsoil acting as the natural foundation cannot be affected by the weather. This depth would normally not be less than 1 m.

The narrow strip foundation is not suitable for buildings with heavy structural loading or where the subsoil is weak in terms of supporting the combined loads imposed on it. Where this is the case, a wide strip foundation is needed.

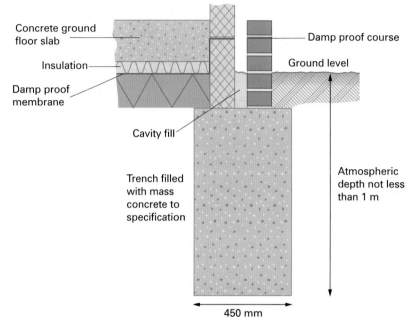

Figure 5.5 Narrow strip foundation

Wide strip foundations

Wide strip foundations consist of steel reinforcement placed within the concrete base of the foundation. This removes the need to increase the depth considerably in order to spread the heavier loads adequately.

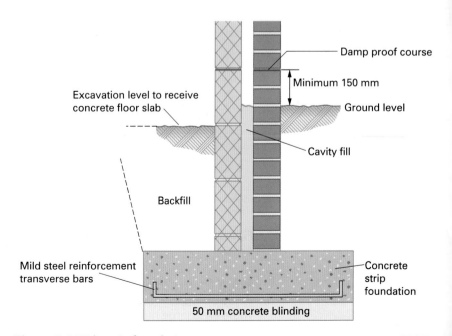

Figure 5.6 Wide strip foundation

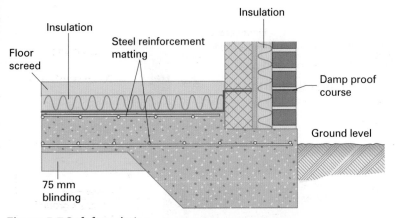

Figure 5.7 Raft foundation

Raft foundations

These types of foundation are used where the soil has poor bearing capacity, making the soil prone to settlement. A raft foundation consists of a slab of reinforced concrete covering the entire base of the structure. The depth of the concrete is greater around the edges of the raft in order to protect the load-bearing soil directly beneath the raft from further effects of moisture taken in from the surrounding area.

Pad foundations

Pad foundations are used where the main loads of a structure are imposed at certain points. An example would be where brick or steel columns support the weight of floors or roof members, and walls between the columns are of non-load-bearing cladding panels. The simplest form of pad foundation is where individual concrete pads are placed at various points around the base of the structure, and concrete ground beams span across them. The individual concrete pads will absorb the main imposed loads, while the beams will help support the walls.

The depth of a pad foundation will depend on the load being imposed on it; in some instances, it may be necessary to use steel reinforcement to prevent excessive depths of concrete. This type of pad foundation can reduce the amount of excavation work required, as trenches do not need to be dug out around the entire base of the proposed structure.

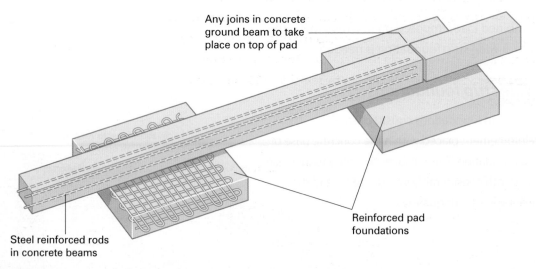

Figure 5.8 Square pad foundation with spanning ground beams

Piled foundations

There are a large number of different types of piled foundations, each with an individual purpose in relation to the type of structure and ground conditions.

Short bored piled foundations are the most common piled foundations. They are predominantly used for domestic buildings where the soil is prone to movement, particularly at depths below 1 m.

A series of holes are bored, by mechanical means, around the perimeter of the base of the proposed building. The diameter of the bored holes will normally be between 250 and 350 mm and can extend to depths of up to 4 m. Once the holes have been bored, shuttering is constructed to form lightweight reinforced concrete beams, which span across the bored piles. The bored holes are then filled with concrete, with reinforcement projecting from the top of the pile concrete, so it can be incorporated into the concrete beams that span the piles.

As with the pad foundation, short bored piled foundations can significantly reduce the amount of excavated soil because there is no need to excavate deep trenches around the perimeter of the proposed structure.

Stepped foundations

A stepped foundation is used on sloping ground. The height of each step should not be greater than the thickness of the concrete, and should not be greater than 450 mm. Where possible, the height of the step should coincide with brick course height in order to avoid oversized mortar bed joints and eliminate the need for split brick courses. The overlap of the concrete to that below should not be less than 300 mm or less than the thickness of the concrete.

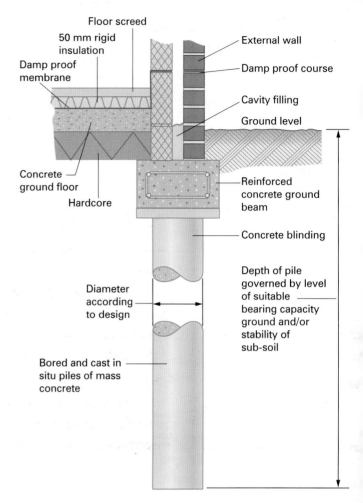

Figure 5.9 Typical short bored piled foundation

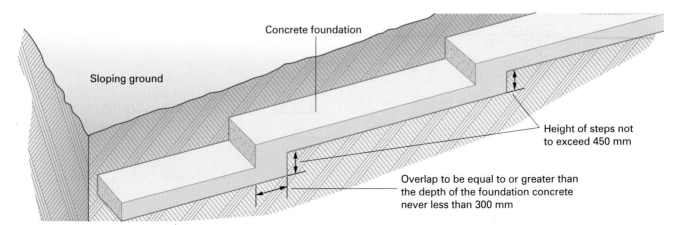

Figure 5.10 Typical stepped foundation

Activity

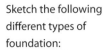

Sketch the following different types of foundation:

- strip foundation
- wide strip foundation
- raft foundation
- pad foundation
- piled foundation
- stepped foundation.

Knowledge refresher

1. What should happen before any construction work commences on a building project?

2. State three factors that influence the design of a foundation.

3. Explain what is meant by a 'dead load'.

4. Explain what is meant by an 'imposed load'.

5. During a site investigation, certain data need to be collected. Name three types of data that have to be recorded.

6. Why must excavation work be carefully planned before it is carried out?

7. Name three categories of soil.

8. Name three types of foundation.

9. How can surface water affect excavation work?

10. In a stepped foundation, what is the recommended maximum height of each concrete step?

What would you do?

1   You have been tasked with building a garage. You decide
     not to go with a soil survey as this is an extra expense, and you just
     put in a standard strip foundation. What could go wrong? What
     could the cost implications be? What other implications could
     there be?

2   You have been asked by your boss to enter an excavation to
     clean out some loose soil. The excavation is 1.5 m deep, and has
     been excavated for a foundation. There are no supports on the
     excavation sides and overnight there has been a considerable
     amount of rainfall. Your boss shouts at you to get on with it as the
     concrete for the foundation will be here in 10 minutes. What should
     you do? What could the implications be if you do it? What could the
     implications be if you don't?

Exterior walls

External walls come in a variety of types, but the most common is cavity walling. Cavity walling is simply two masonry walls built parallel to each other, with a gap between acting as the cavity. The cavity wall acts as a barrier to weather, with the outer leaf preventing rain and wind penetrating the inner leaf. The cavity is usually filled with insulation to prevent heat loss.

How cavity walls are constructed

Cavity walls mainly consist of a brick outer skin and a blockwork inner skin. There are instances where the outer skin may be made of block and then rendered or covered by tile hanging. The minimum cavity size allowed is 75 mm but the cavity size is normally governed by the type and thickness of insulation to be used and whether the cavity is to be fully filled or partially filled with insulation.

The thickness of blocks used will also govern the overall size of the cavity wall. On older properties, the internal blocks were always of 100 mm thickness. Nowadays, due to the emphasis on energy conservation and efficiency, blocks are more likely to be 125 mm or more.

In all cases, the cavity size will be set out to the drawing with overall measurements specified by the architect and to local authority requirements.

Once the **foundations** have been concreted the **sub-structural walling** can be constructed, usually by using blocks for both walls (see Figure 5.11).

Definition

Foundations – concrete bases supporting walls

Sub-structural walling – brickwork between the foundation concrete and the horizontal damp proof course (DPC)

Remember

The correct size must be used for the internal wall, with the cavity size to suit.

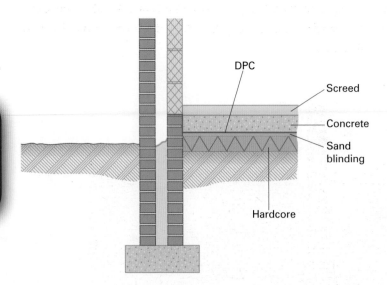

Figure 5.11 Section of sub-structural walling

In some situations trench blocks may be used below ground level and then traditional cavity work constructed up to the damp proof course (DPC). A horizontal DPC must be inserted at a minimum height of 150 mm above ground level to both walls. This is to prevent damp rising, below ground, up through the block and brickwork to penetrate to the inside. The cavity must also be filled with weak concrete to ground level to help the sub-structural walling resist lateral pressure.

Cavity walls above DPC

The older, traditional way to build a cavity wall is to build the brickwork first and then the blockwork. Now, due to the introduction of insulation into the cavity, the blockwork is generally built first, especially when the cavity is partially filled with insulation. This is because the insulation requires holding in place against the internal block wall, by means of special clips that are attached to the **wall ties**. In most cases the clips are made of plastic as they do not rust or rot. The reason for clipping the insulation is to stop it from moving away from the blocks, which would cause the loss of warmth to the interior of the building, as well as causing a possible **bridge** of the cavity, which could cause a damp problem.

The brick courses should be gauged at 75 mm per course but sometimes course sizes may change slightly to accommodate window or door heights. In most instances these positions and measurements are designed to work to the standard gauge size. This will also allow the blockwork to run level at every third course of brick, although the main reason will be explained in the wall tie section below.

On most large sites, patent types of corner profile are used rather than building traditional corners (see Figure 5.12). These allow the brickwork to be built faster and, if set up correctly, more accurately. But they must also be marked for the gauge accurately and it makes sense to mark window sill heights or **window heads** and door heights so they do not get missed, which would result in brickwork being taken down.

Activity

Sketch a section through a cavity wall including DPC and wall ties.

Definition

Wall ties – stainless steel or plastic fixings to tie cavity walls together

Bridge – where moisture can be transferred from the outer wall to the inner leaf by material touching both walls

Window head – top of a window

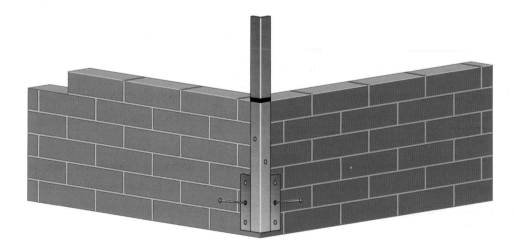

Figure 5.12 A corner profile set up

Wall ties

Wall ties are a very important part of a cavity wall as they tie the internal and external walls together, resulting in a stronger job. If we built cavity walls to any great height without connecting them together, the walls would be very unstable and could possibly collapse.

A wall tie should be:

1. rust-proof
2. rot-proof
3. of sufficient strength
4. able to resist moisture.

There are many designs of wall tie currently on the market, with a wide selection suitable for all types of construction methods. One of the most common types used when tying together brick and block leaves is the masonry general purpose tie. These ties are made from very strong stainless steel, and incorporate a twist in the steel at the mid point of the length. This twist forms a drip system, which prevents the passage of water from the outer to the inner leaf of the structure.

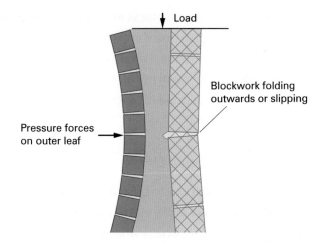

Figure 5.13 Section of wall without wall ties

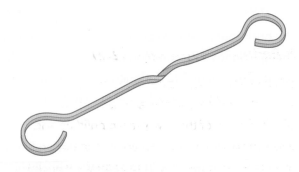

Figure 5.14 General purpose wall tie

You must take care to keep the wall ties clean when they are placed in the wall: if bridging occurs, it may result in moisture penetrating the internal wall.

The positioning and density of wall ties

In cavity walling where both the outer and inner leaves are 90 mm or thicker, you should use ties at not less than 2.5 per square metre, with 900 mm maximum horizontal distance by a maximum 450 mm vertical distance and staggered.

At positions such as vertical edges of an opening, unreturned or unbonded edges and vertical expansion joints, you need to use additional ties at a maximum of 300 mm in height (usually 225 mm to suit block course height) and located not more than 225 mm from the edge. Wall ties should be bedded into each skin of the cavity wall to a minimum distance of 50 mm.

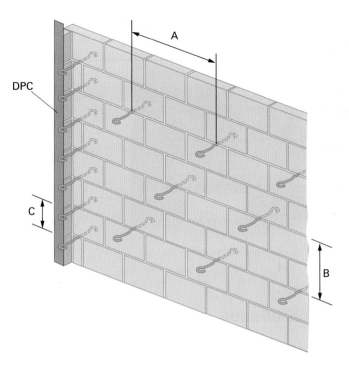

DPC

A

C

B

A 900 mm maximum horizontal distance

B 450 mm maximum vertical distance

C Additional ties, 300 mm maximum vertical distance

Figure 5.15 Spacing of wall ties

Did you know?

Any batten can be used as long as the width is the same as the cavity space.

Definition

Cavity batten – a timber piece laid in a cavity to prevent mortar droppings falling down the cavity

Keeping a cavity wall clean

It is important to keep the cavity clean to prevent dampness. If mortar is allowed to fall to the bottom of the cavity it can build up and allow the damp to cross and enter the building. Mortar can also become lodged on the wall ties and create a bridge for moisture to cross. We can prevent this by the use of **cavity battens**, pieces of timber the thickness of the cavity laid on to the wall ties and attached by wires or string (to prevent dropping down the cavity) to the wall and lifted alternately as the wall progresses.

The bottom of the wall can be kept clean by either leaving bricks out or bedding bricks with sand so they can be taken out to clean the cavity. These are called core holes and are situated every fourth brick along the wall to make it easy to clean out each day. Once the wall is completed the bricks are bedded into place.

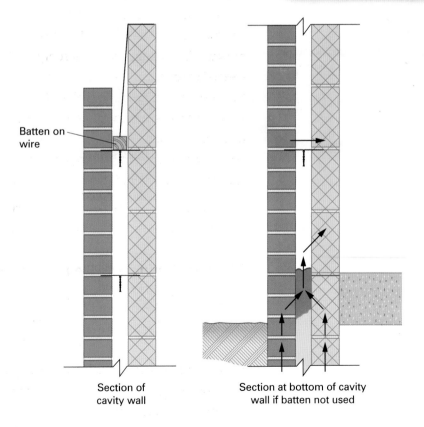

Batten on wire

Section of cavity wall

Section at bottom of cavity wall if batten not used

Figure 5.16 Cavity batten in use

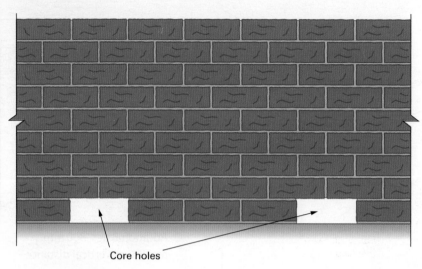

Core holes

Elevation of wall

Figure 5.17 Core holes

Steps to take to prevent damp penetration

- Set out openings carefully to avoid awkward bonds.
- Care is needed in construction to make sure dampness or water does not enter the building.
- DPCs and wall ties should be carefully positioned.
- Steel cavity lintels should have minimum 150 mm bearings solidly bedded in the correct position.
- Weep holes should be put in at 450 mm centres immediately above the lintel in the outer leaf.

No insulation has been shown in the drawings because they only show one situation. In most cavity wall construction, insulation of one kind or another will have to be incorporated to satisfy current *Building Regulations*.

Fire spread

In addition to the prevention of damp penetration and cold bridging, there is a requirement under the *Building Regulations* that cavities and concealed spaces in a structure or fabric of a building are sealed by using cavity barriers or fire stopping. This cuts down the hidden spread of smoke or flames in the event of fire breaking out in a building.

Closing at eaves level

The cavity walls have to be 'closed off' at roof level for two main reasons:

1. To prevent heat loss and the spread of fire.

2. To prevent birds or vermin entering and nesting.

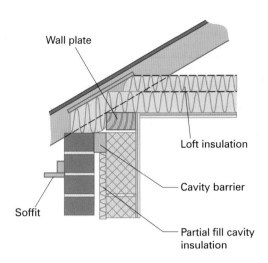

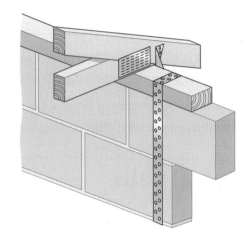

Figure 5.18 Roof section

This area of the wall is where the roof is connected, by means of a timber wall plate bedded on to the inner leaf. The plate is then secured by means of restraint straps that are galvanised 'L' shaped straps screwed to the top of the wall plate and down the blockwork. This holds the roof structure firmly in place and also prevents the roof from spreading under the weight of the tiles, etc. The minimum distance that the straps should be apart is 1.2 m. In some instances they may be connected directly from the roof truss to the wall.

If a gable wall is required, restraint straps should be used to secure the roof to the end wall (see Figure 5.18).

The external wall can be built to the height of the top of the truss so as not to leave gaps, or 'closed off' by building blocks laid flat to cover the cavity above the external soffit line from inside, avoiding damp penetration. In some instances the cavity may be left open with the cavity insulation used as the seal.

Timber-kit houses

Timber-kit houses are becoming more and more popular as they can be erected to a windtight and watertight stage within a few days. The principle is similar to a cavity wall, but here the inner skin is a timber frame, which is clad in timber sheet material and covered in a breathable membrane, to prevent water and moisture penetrating the timber. The outer skin is usually face brickwork.

Constructing timber-kit houses

Timber-kit house construction starts off in exactly the same way as any house build, with the foundations and the cavity wall built up to DPC level. However, from then on, the construction method is completely

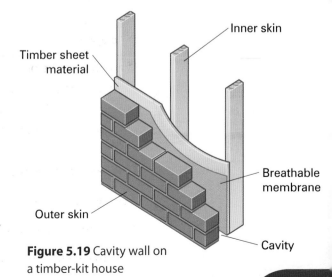

Figure 5.19 Cavity wall on a timber-kit house

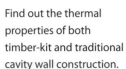
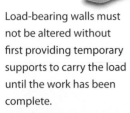
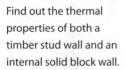
different, as the interior skin of the cavity is timber framed as opposed to block work. The timber-kit house is created in panels within a factory environment and delivered in sections on the back of a lorry.

The timber panels are lifted into place (usually by crane) and are bolted together. Once the wall panels are in place, the exterior face brickwork can begin.

There are also other types of exterior walling, such as solid stone or log cabin style. Industrial buildings may have steel walls clad in sheet metal.

Internal walls

Internal walls are either load bearing – meaning they support any upper floors or roof – or are non-load-bearing, used to divide the floor space into rooms.

Internal walls also come in a variety of styles. Here is a list of the most common types.

- **Solid block walls** – simple block work, either covered with plasterboard or plastered over to give a smooth finish, to which wallpaper or paint is applied. Solid block walls offer low thermal and sound insulation qualities, but advances in technology and materials means that blocks manufactured from a lightweight aggregate can give better sound and heat insulation.

- **Solid brick walling** – usually made with face brickwork as a decorative finish. It is unusual for all walls within a house to be made from brickwork.

- **Timber stud walling** – more common in timber kit houses and newer buildings. Timber stud walling is also preferred when dividing an existing room, as it is quicker to erect. Clad in plasterboard and plastered to a smooth finish, timber stud partitions can be made more fire resistant and sound/thermal qualities can be improved with the addition of insulation or different types of plasterboard. Another benefit of timber stud walling is that timber noggins can be placed within the stud to give additional fixings for components such as radiators or wall units. Timber stud walling can also be load bearing, in which case thicker timbers are used.

- **Metal stud walling** – similar to timber stud, except metal studs are used and the plasterboard is screwed to the studding.

- **Grounds lats** – timber battens that are fixed to a concrete or stone wall to provide a flat surface, to which plasterboard is attached and a plaster finish applied

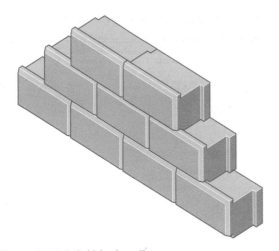

Figure 5.20 Solid block wall

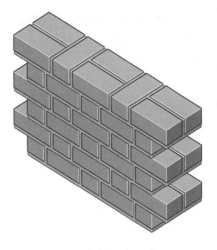

Figure 5.21 Solid brick wall

Figure 5.22 Timber stud wall

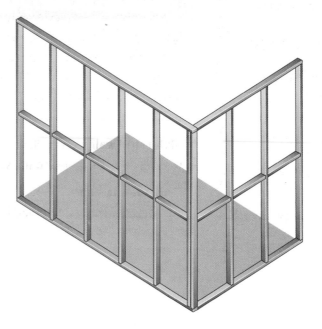

Figure 5.23 Metal stud wall

Knowledge refresher

1 What is the minimum cavity size for a cavity wall?

2 What is the purpose of wall ties in a cavity wall?

3 State one advantage of using a corner profile.

4 What is the main reason a cavity needs to be kept clean?

5 Outline three steps to take to prevent damp penetration in a cavity wall.

6 Give two reasons why cavity walls have to be closed off at roof level.

7 State one advantage of using timber kit as opposed to cavity walling.

8 How are timber panels in a kit house joined together?

9 How could a timber stud wall be made more soundproof?

10 Describe how a solid internal block wall can be finished.

What would you do?

1 You are going to build your own house. Using the information you gained from previous Activities and what you know, what type of house should you build: cavity or kit?

2 You have been tasked with creating an opening in an internal block work wall. What can you do to check if the wall is load bearing? What should be done if the wall is load bearing? What could the consequences be if the wall is load bearing and you create the opening without shoring?

Floors

Floors have a number of standard components, including the following:

- **DPC**

This is the damp proof course that is inserted into both skins of the external cavity wall construction. It should be a proprietary product that is tested and has a long life expectancy.

- **DPM**

This is the damp proof membrane, which is placed in large sheets within the floor structure so it can resist the passage of moisture and rising damp. This keeps the floors dry. The DPM should be taken up vertically and tucked into the DPC to form a complete seal. All DPM should be lapped by at least 300 mm, with any joints taped.

- **Screeds**

Floor screeds are considered in the solid concrete floor section (see page 118). They provide a finish to the concrete surface, cover up services and provide a level for floor finishes to be applied to. They also provide falls to floors: for example, in a wet room for the shower waste.

- **Wall plate**

The wall plate is on top of the sleeper wall that supports the floor joists. It has a DPC underneath it to prevent damp penetrating the timber. Wall plates should be treated.

The most common types of flooring used are:

- suspended timber floors
- solid concrete floors
- pre-cast beam floors
- floating floors.

Suspended timber floors

Suspended timber floors can be fitted at any level, from top floor to ground floor. In the next few pages, you will look at:

- basic structure
- joists
- construction methods
- floor coverings.

Basic structure

Suspended timber floors are constructed with timbers known as joists, which are spaced parallel to each other spanning the distance of the building. Suspended timber floors are similar to traditional roofs in that they can be single or double, a single floor being supported at the two ends only and a double floor supported at the two ends and in the middle by way of a honeycombed sleeper/dwarf wall, steel beam or load-bearing partition.

All floors must be constructed to comply with the *Building Regulations*, in particular Part C, which is concerned with damp. The bricklayer must insert a **damp proof course (DPC)** between the brick or block work when building the walls, situated no less than 150 mm above ground level. This prevents moisture moving from the ground to the upper side of the floor. No timbers are allowed below the DPC. Air bricks, which are built into the external walls of the building, allow air to circulate round the underfloor area, keeping the moisture content below the dry rot limit of 20 per cent, thus preventing dry rot.

Definition

Damp proof course (DPC) – a substance that is used to prevent damp from penetrating a building

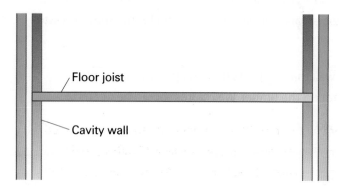

Figure 5.24 Single floor

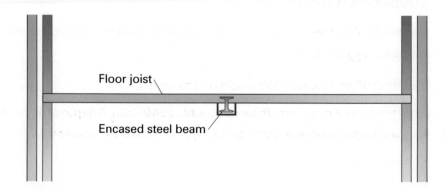

Figure 5.25 Double floor

Joists

In domestic dwellings suspended upper floors are usually single floors, with the joist supported at each end by the structural walls but, if support is required, a load-bearing partition is used. The joists that span from one side of the building to the other are called bridging joists, but any joists that are affected by an opening in the floor such as a stairwell or chimney are called trimmer, trimming and trimmed joists.

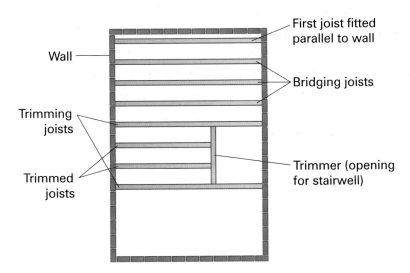

Figure 5.26 Joists and trimmers

Bridging joists are usually sawn timber 50 mm thick. The trimmer that carries the trimmed joists and transfers this load to the trimming must be thicker – usually 75 mm sawn timber, or in some instances two 50 mm bridging joists bolted together. The depth of the joist is easily worked out by using the calculation:

Depth of joist = span / 20 + 20

So, for example, if you have a span of 4 m

Depth = 4000/20 + 20

Depth = 200 + 20

Depth = 220

the depth of the joist required would be 220 mm.

If the span was 8 m, the depth would double to 440 mm. A depth of 440 mm is too great, so you would need to look at putting in a support to create a double floor.

Types of joist

As well as the traditional method of using solid timber joists, there are now alternatives available. These are the most common.

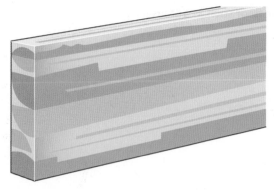

Figure 5.27 Laminated joist

Laminated joists

These were originally used for spanning large distances, as a laminated beam could be made to any size, but now they are more commonly used as an environmental alternative to solid timber – recycled timber can be used in the laminating process. They are more expensive than solid timber, as the joists have to be manufactured.

I type joists

These are now some of the most commonly used joists in the construction industry: they are particularly popular in new build and are the only joists used in timber kit house construction. I type beam joists are lighter and more environmentally friendly as they use a composite panel in the centre, usually made from oriented strand board, which can be made from recycled timber.

The following construction method shows how to fit solid timber joists but whichever joists you use, the methods are the same.

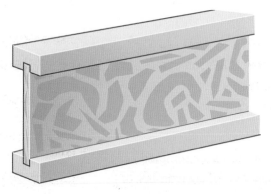

Figure 5.28 I type joist

Construction methods

A suspended timber floor must be supported either end. The figures below show ways of doing this.

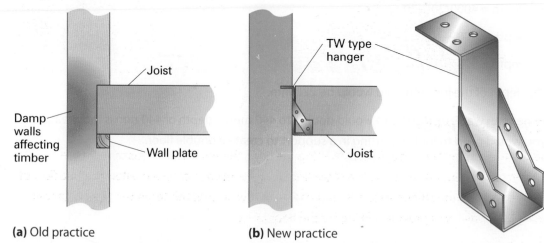

(a) Old practice

(b) New practice

Figure 5.29 Solid floor bearings

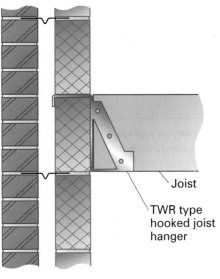

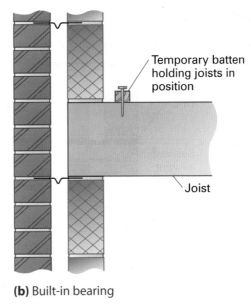

(a) Joist hanger bearing

(b) Built-in bearing

Figure 5.30 Cavity wall bearings

If a timber floor has to trim an opening, there must be a joint between the trimming and the trimmer joists. Traditionally, a **tusk tenon joint** was used (even now, this is sometimes preferred) between the trimming and the trimmer joist. If the joint is formed correctly a tusk tenon is extremely strong, but making one is time-consuming. A more modern method is to use a metal framing anchor or timber-to-timber joist hanger.

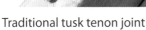

Traditional tusk tenon joint

Joist hanger

Fitting floor joists

Before the carpenter can begin constructing the floor, the bricklayer needs to build the honeycomb sleeper walls. This type of walling has gaps in each course to allow the free flow of air through the underfloor area. It is on these sleeper walls that the carpenter lays his timber wall plate, which will provide a fixing for the floor joists.

The following pages describe the steps in fitting floor joists.

Activity

Sketch a section through a suspended timber floor, showing how the flooring is supported.

Definition

Tusk tenon joint – a kind of mortise and tenon joint that uses a wedge-shaped key to hold the joint together

Activity

With the aid of a sketch, explain the purpose of a joist hanger and how it functions.

Activity

Sketch a tusk tenon joint.

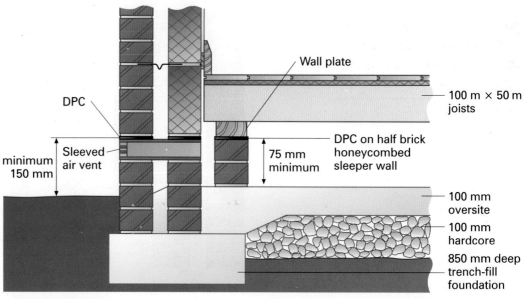

DPC

Sleeved air vent

minimum 150 mm

Wall plate

100 m × 50 m joists

DPC on half brick honeycombed sleeper wall

75 mm minimum

100 mm oversite

100 mm hardcore

850 mm deep trench-fill foundation

Figure 5.31 Section through floor and wall

Step 1 Bed and level the wall plate onto the sleeper wall with the DPC under it.

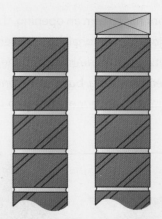

Figure 5.32 Step 1 Bed in the wall plate

Step 2 Cut joists to length and seal the ends with a coloured preservative. Mark out the wall plate with the required centres, space the joists out and fix temporary battens near each end to hold the joists in position. Ends should be kept away from walls by approximately 12 mm. It is important to ensure that the camber is turned upwards.

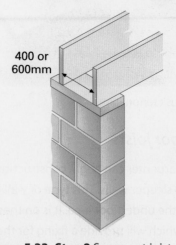

400 or 600mm

Figure 5.33 Step 2 Space out joists

Step 3 Fix the first joist parallel to the wall with a gap of 50 mm. Fix trimming and trimmer joists next to maintain the accuracy of the opening.

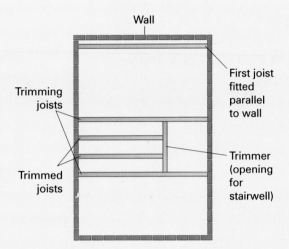

Figure 5.34 Step 3 Fit first joist and trimmers

Step 4 Fix subsequent joists at the required spacing as far as the opposite wall. Spacing will depend on the size of joist and/or floor covering, but usually 400 mm to 600 mm centres are used.

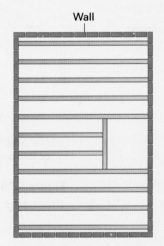

Figure 5.35 Step 4 Fit remaining joists

Step 5 Fit folding wedges to keep the end joists parallel to the wall. Overtightening is to be avoided in case the wall is strained.

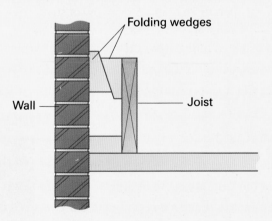

Figure 5.36 Step 5 Fit folding wedges

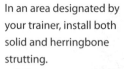
Step 6 Check that the joists are level with a straight edge or line and, if necessary, pack with slate or DPC.

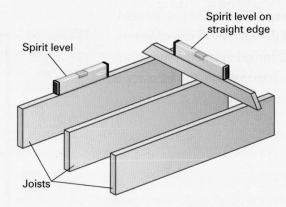

Figure 5.37 Step 6 Ensure joists are level

Step 7 Fit restraining straps and, if the joists span more than 3.5 m, fit strutting and bridging, described in more detail next.

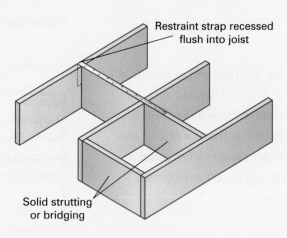

Figure 5.38 Step 7 Fix restraining straps, struts and bridges

Strutting and bridging

When joists span more than 3.5 m, a row of struts must be fixed midway between each joist. Strutting or bridging stiffens the floor in the same way that noggins stiffen timber stud partitions, preventing movement and twisting, which is useful when fitting flooring and ceiling covering. A number of methods are used, but the main ones are solid bridging, herringbone strutting and steel strutting.

Solid bridging

For solid bridging, timber struts the same depth as the joists are cut to fit tightly between each joist and **skew-nailed** in place. A disadvantage of solid bridging is that it tends to loosen when the joists shrink.

Solid bridging

Herringbone strutting

Here timber battens (usually 50 × 25 mm) are cut to fit diagonally between the joists. A small saw cut is put into the ends of the battens before nailing to avoid the battens splitting. This will remain tight even after joist shrinkage. The following steps describe the fitting of timber herringbone strutting.

Step 1 Nail a temporary batten near the line of strutting to keep the joists spaced at the correct centres.

Space joists

Step 2 Mark the depth of a joist across the edge of the two joists, then measure 12 mm inside one of the lines and remark the joists. The 12 mm less than the depth of the joist ensures that the struts will finish just below the floor and ceiling level (as shown in step 5).

Mark joist depths

Step 3 Lay the strut across two joists at a diagonal to the lines drawn in Step 2.

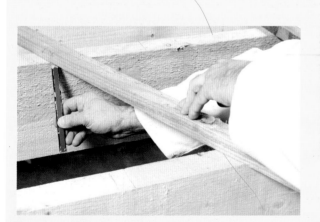

Lay struts across two joists at a diagonal

Step 4 Draw a pencil line underneath as shown in Step 3 and cut to the mark. This will provide the correct angle for nailing.

Cut to the mark

Step 5 Fix the strut between the two joists. The struts should finish just below the floor and ceiling level. This prevents the struts from interfering with the floor and ceiling if movement occurs.

Fix the strut

Steel strutting

There are two types of galvanised steel herringbone struts available.

The first has angled lugs for fixing with the minimum 38 mm round head wire nails.

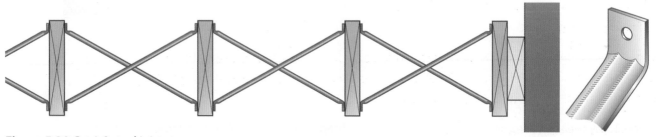

Figure 5.39 Catric® steel joist struts

The second has pointed ends, which bed themselves into joists when forced in at the bottom and pulled down at the top. Unlike other types of strutting, this type is best fixed from below.

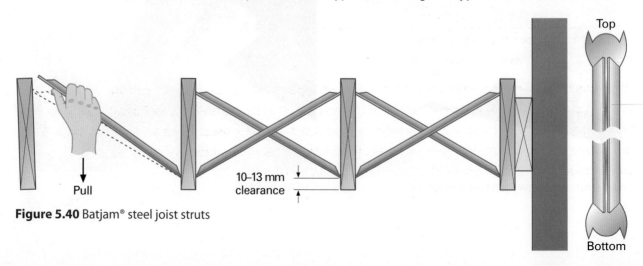

Pull

10–13 mm clearance

Top

Bottom

Figure 5.40 Batjam® steel joist struts

The disadvantage of steel strutting is that it only comes in set sizes, to fit centres of 400, 450 and 600 mm. This is a disadvantage as there will always be a space in the construction of a floor that is smaller than the required centres.

Restraint straps

Anchoring straps, normally referred to as restraint straps, are needed to restrict any possible movement of the floor and walls due to wind pressure. They are made from galvanised steel, 5 mm thick for horizontal restraints and 2.5 mm for vertical restraints, 30 mm wide and up to 1.2 m in length. Holes are punched along the length to provide fixing points.

When the joists run parallel to the walls, the straps will need to be housed into the joist to allow the strap to sit flush with the top of the joist, keeping the floor even. The anchors should be fixed at a maximum of 2 m centre to centre. More information can be found in schedule 7 of the *Building Regulations*.

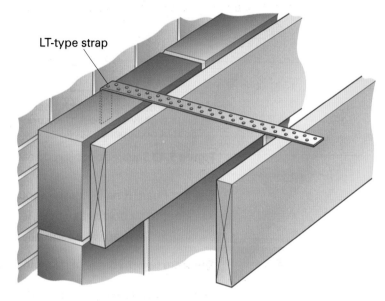

Figure 5.41 Restraint straps for joists parallel or at right angles to a wall

Floor coverings

Softwood flooring

Softwood flooring can be used at either ground or upper floor levels. It usually consists of 25 × 150 mm tongued and grooved (T&G) boards. The tongue is slightly off centre to provide extra wear on the surface that will be walked upon.

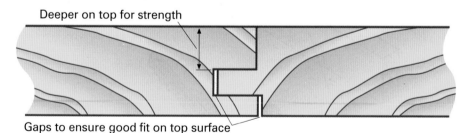

Figure 5.42 Section through softwood covering

When boards are joined together, the joints should be staggered evenly throughout the floor to give it strength. They should never be placed next to each other, as this prevents the joists from being tied together properly. The boards are either fixed with floor brads nailed through the surface and punched below flush, or secret nailed with lost head nails through the tongue. The nails used should be 2½ times the thickness of the floorboard.

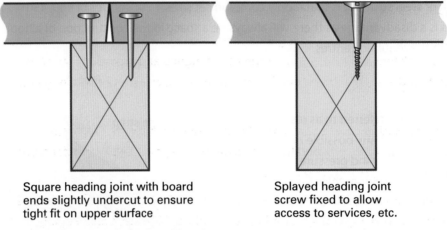

Square heading joint with board ends slightly undercut to ensure tight fit on upper surface

Splayed heading joint screw fixed to allow access to services, etc.

Figure 5.43 Square and splayed heading

The first board is nailed down about 10–12 mm from the wall. The remaining boards can be fixed four to six boards at a time, leaving a 10–12 mm gap around the perimeter to allow for expansion. This gap will eventually be covered by the skirting board.

There are two methods of clamping the boards before fixing:

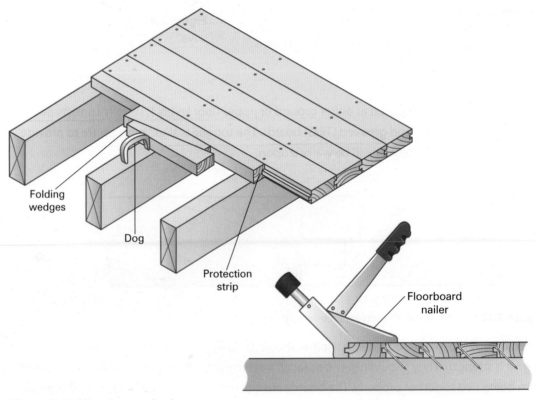

Folding wedges

Dog

Protection strip

Floorboard nailer

Figure 5.44 Clamping methods

Chipboard flooring

Flooring-grade chipboard is increasingly being used for domestic floors. It is available in sheets sizes of 2440 × 600 × 18 mm and can be square edged or tongued and grooved on all edges, the latter being preferred. If square-edged chipboard is used it must be supported along every joint.

Tongued and grooved boards are laid end to end, at right angles to the joists. Cross-joints should be staggered and, as with softwood flooring, expansion gaps of 10–12 mm left around the perimeter. The ends must be supported.

When setting out the floor joists, the spacing should be set to avoid any unnecessary wastage. The boards should be glued along all joints and fixed using either 50–65 mm annular ring shank nails or 50–65 mm screws. Access traps must be created in the flooring to allow access to services such as gas and water.

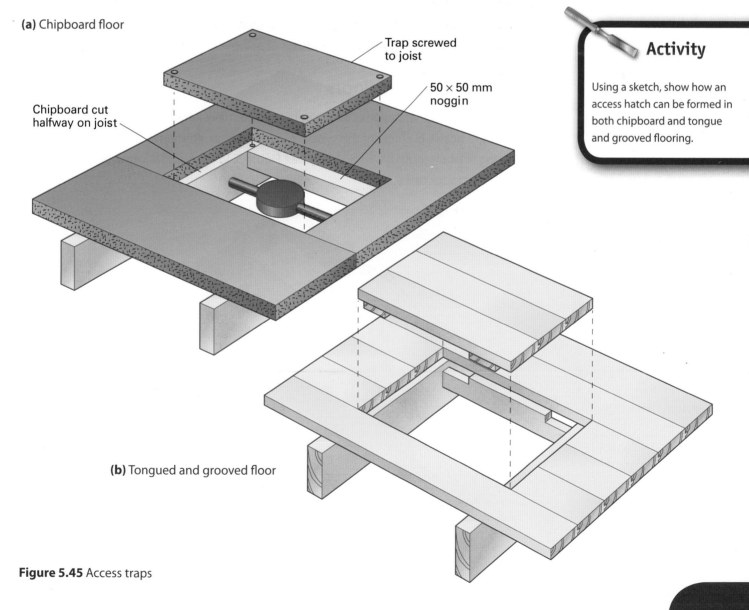

(a) Chipboard floor

Trap screwed to joist

50 × 50 mm noggin

Chipboard cut halfway on joist

(b) Tongued and grooved floor

Figure 5.45 Access traps

Activity

Using a sketch, show how an access hatch can be formed in both chipboard and tongue and grooved flooring.

Solid concrete floors

Solid concrete floors are more durable than suspended timber floors. They are constructed on a sub-base incorporating hardcore, damp proof membranes and insulation. The depth of the hardcore and concrete will depend on the nature of the building, and will be set by the Building Regulations and the local authority.

In the next few pages, you will look at:

- formwork for concrete floors
- reinforcement
- compacting of concrete
- surface finishes
- curing.

Any concreting job has to be supported at the sides to prevent the concrete just running off and this support comes in the way of formwork.

Formwork for ground floors

Floors for buildings such as factories and warehouses etc. have large areas and would be difficult to lay in one slab. Floors of this type are usually laid in alternative strips up to 4.5 m wide, running the full length of the building (see Figure 5.46). The actual formwork would be similar to that used for paths.

Activity

Using a sketch, show how formwork for ground floors can be created.

Activity

In an area designated by your trainer, create formwork for concrete flooring.

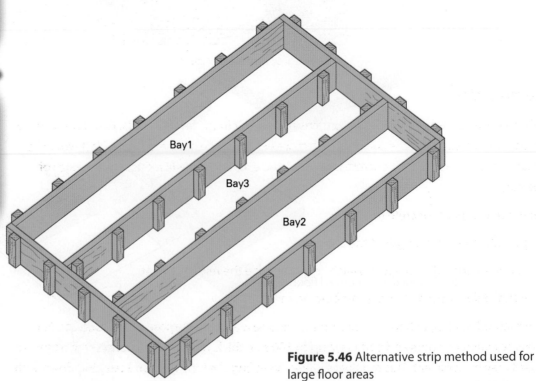

Bay1

Bay3

Bay2

Figure 5.46 Alternative strip method used for large floor areas

Reinforcement

Concrete is strong in **compression** but weak in **tension** so, to prevent concrete from being 'pulled' apart when under pressure, steel reinforcement is provided. The type and position of the reinforcement will be specified by the structural engineer.

The reinforcement must always have a suitable thickness of concrete cover to prevent the steel from rusting if exposed to moisture or air. The amount of cover required depends upon the location of the site with respect to exposure conditions, and ranges from 20 mm in mild exposure to 60 mm for very severe exposure to water.

Definition

Compression – being squeezed or squashed together

Tension – being stretched

To prevent the reinforcement from touching the formwork, spacers made from concrete, fibre cement or plastic are used. They are available in several shapes and various sizes to give the correct cover.

Steel reinforcement of concrete

Compacting

When concrete has been placed, it contains trapped air in the form of voids. To get rid of these voids we must compact the concrete. The more workable the concrete the easier it would be to compact, but also if the concrete is too wet, the excess water will reduce the strength of the concrete.

Failure to compact concrete results in:

- reduction in the strength of the concrete
- water entering the concrete, which could damage the reinforcement
- visual defects, such as honeycombing on the surface.

The method of compaction depends on the thickness and the purpose of the concrete. For oversite concrete, floors and pathways up to 100 mm thick, manual compaction with a tamper board may be sufficient. This requires slightly overfilling the formwork and tamping down with the tamper board. For larger spans the tamper board may be fitted with handles.

Tamper board with handles

For slabs up to 150 mm thick, a vibrating beam tamper should be used. This is simply a tamper board with a petrol-driven vibrating unit bolted on. The beam is laid on the concrete with its motor running and is pulled along the slab.

For deeper structures, such as retaining walls for example, a poker vibrator would be required. The poker vibrator is a vibrating tube at the end of a flexible drive connected to a petrol motor. The pokers are available in various diameters from 25 mm to 75 mm.

The concrete should be laid in layers of 600 mm with the poker in vertically and penetrating the layer below by 100 mm. The concrete is vibrated until the air bubbles stop and the poker is then lifted slowly and placed 150 to 1000 mm from this incision, depending on the diameter of the vibrator.

Surface finishes

Vibrating beam tamper

Vibrating poker in use

Surface finishes for slabs may be:

- **Tamped finish.** Simply using a straight edge or tamper board when compacting the concrete will leave a rough finish to the floor, ideal for a path or drive surface, giving grip to vehicles and pedestrians. This finish may also be used if a further layer is to be applied to give a good bond.

- **Float and brush finish.** After **screeding** off the concrete with a straight edge, the surface is floated off using a steel or wooden float and then brushed lightly with a soft brush (see photo below). Again, this would be suitable for pathways and drives.

- **Steel float finish.** After screeding off using a straight edge, a steel float is applied to the surface. This finish attracts particles of cement to the surface, causing the concrete to become impermeable to water but also very slippery when wet. This is not very suitable for outside but ideal for use indoors for floors, etc.

- **Power trowelling/float.** Three hours after laying, a power float is applied to the surface of the concrete. After a further delay to allow surface water to evaporate, a power trowel is then used. A power float has a rotating circular disc or four large flat blades powered by a petrol engine. The edges of the blades are turned up to prevent them digging into the concrete slab. This finish would most likely be used in factories where a large floor area would be needed.

- **Power grinding.** This is a technique used to provide a durable wearing surface without further treatment. The concrete is laid, compacted and trowel finished. After 1 to 7 days the floor is ground, removing the top 1–2 mm, leaving a polished concrete surface.

Definition

Screeding – levelling off concrete by adding a final layer

Remember

Make sure you always clean all tampers and tools after use.

Did you know?

The success of surface finishes depends largely on timing. You need to be aware of the setting times in order to apply the finish.

Brushed concrete finish

Power float

Surface treatment for other surfaces may be:

- **Plain smooth surfaces.** After the formwork has been struck, the concrete may be polished with a carborundum stone, giving a polished water-resistant finish.

- **Textured and profiled finish.** A simple textured finish may be made by using rough sawn boards to make the formwork. When struck, the concrete takes on the texture of these boards. A profiled finish can be made by using a lining inside the formwork. The linings may be made from polystyrene or flexible rubber-like plastics, and gives a pattern to the finished concrete.

- **Ribbed finish.** These are made by fixing timber battens to the formwork.

Ribbed concrete finish

- **Exposed aggregate finish.** The coarse aggregate is exposed by removing the sand and cement from the finished concrete with a sand blaster. Another method of producing this finish is by applying a chemical retarder to the formwork, which prevents the cement in contact with it from hardening. When the formwork is removed, the mortar is brushed away to uncover the aggregate in the hardened concrete.

Curing

When concrete is mixed, the quantity of water is accurately added to allow for hydration to take place. The longer we can keep this chemical reaction going, the stronger the concrete will become.

To allow the concrete to achieve its maximum strength, the chemical reaction must be allowed to keep going for as long as possible. To do this we must 'cure' the concrete. This is done by keeping the concrete damp and preventing it from drying out too quickly.

Curing can be done by:

1. Spraying the concrete with a chemical sealer, which dries to leave a film of resin to seal the surface and reduces the loss of moisture.

2. Spraying the concrete with water, which replaces any lost water and keeps the concrete damp. This can also be done by placing sand or hessian cloth or other similar material on the concrete and dampening.

3. Covering the concrete with a plastic sheet or building paper, preventing wind and sun from evaporating the water into the air. Any evaporated moisture due to the heat will condense on the polythene and drip back on to the concrete surface.

Concreting in hot weather

When concreting in temperatures over 20°C, there is a reduction in workability due to the water being lost through evaporation. The cement also tends to react more quickly with water, causing the concrete to set rapidly.

To remedy the problem of the concrete setting quickly, a 'retarding mixture' may be used. This slows down the initial reaction between the cement and water, allowing the concrete to remain workable for longer.

Extra water may be added at the time of mixing so that the workability would be correct at the time of placing.

Water must *not* be added during the placing of the concrete, to make it more workable, after the initial set has taken place in the concrete.

Concreting in cold weather

Water expands when freezing. This can cause permanent damage if the concrete is allowed to freeze when freshly laid or in hardened concrete that has not reached enough strength (5 N/mm², which takes 48 hours).

Concreting should not take place when the temperature is 2°C or less. If the temperature is only slightly above 2°C, mixing water should be heated.

After being laid, the concrete should be kept warm by covering with insulating quilts, which allows the cement to continue its reaction with the water and prevents it from freezing.

Pre-cast beam floors

Solid concrete floors can only be used at ground floor level, but pre-cast beam floors can be fitted at higher levels. They are mostly used in large industrial buildings or blocks of flats.

The construction of these floors is simple, with pre-stressed beams spanning supporting walls, with an infill of pre-cast lightweight concrete filler blocks.

The insulation can be slung underneath the beams or over the blocks, where it is covered with a structural topping. Airbricks are used to ventilate the airspace below the floor.

Safety tip

The beams in a beam and block floor can be heavy, so you may need a crane to position them on the external walls safely.

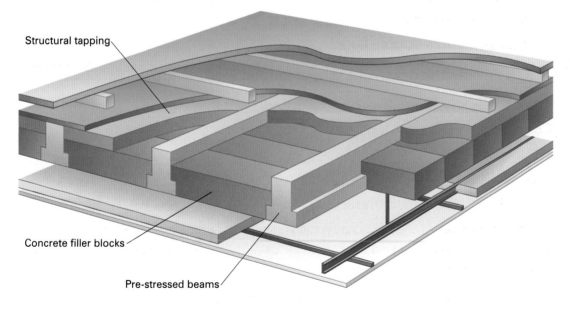

Structural tapping

Concrete filler blocks

Pre-stressed beams

Figure 5.47 Typical beam and block floor

Floating floors

These are basic timber floor constructions that are laid on a solid concrete floor. The timbers are laid in a similar way to joists, although they are usually 50 mm thick maximum as there is no need for support. The timbers are laid on the floor at predetermined centres, and are not fixed to the concrete base (hence floating floor); the decking is then fixed on the timbers. Insulation or underfloor heating can be placed between the timbers to enhance the thermal and acoustic properties.

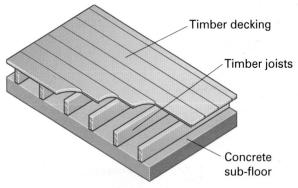

Timber decking

Timber joists

Concrete sub-floor

Figure 5.48 Floating floor

This type of floor 'floats' on a cushion of insulation. Floating floors are normally manufactured from chipboard – either standard or moisture-resistant – for use in bathrooms. This type of floor is ideal for refurbishment work and where insulation upgrading is required.

Activity

Draw a sketch to show how floating floors are created.

Knowledge refresher

1   List the four most common types of floor construction.

2   On a suspended timber floor, what is the difference between a single and a double floor?

3   How can the depth of a timber joist be worked out?

4   What is the difference between solid and herringbone strutting?

5   State the purpose of formwork in concrete floors.

6   Why is reinforcement used in solid concrete flooring?

7   Why is concrete compacted?

8   Describe three ways of finishing a concrete floor.

9   Where would a floating floor be laid?

What would you do?

1   You have been asked to quote for building a garage. The client is unsure of what type of floor to have. They ask your opinion. What type of floor would you select? Why would you select that type of floor? What could influence your selection?

2   You have been tasked with joisting an upper floor. You fit the herringbone strutting but, when it comes to fitting the floor and plasterboarding the ceiling, the strutting is interfering with both the floor and the ceiling. What could have caused this? What could be done to rectify it?

Roofing

Although there are several different types of roofing, all roofs will technically be either a flat roof or a pitched roof.

Flat roofs

A flat roof is any roof which has its upper surface inclined at an angle (also known as the fall, slope or pitch) not exceeding 10 degrees.

A flat roof has a fall to allow rainwater to run off, preventing puddles forming as they can put extra weight on the roof and cause leaks. Flat roofs will eventually leak, so most are guaranteed for only 10 years (every 10 years or so the roof will have to be stripped back and re-covered). Today **fibreglass** flat roofs are available that last much longer, so some companies will give a 25-year guarantee on their roof. Installing a fibreglass roof is a job for specialist roofers.

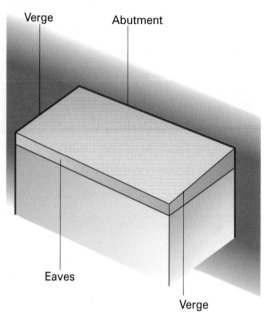

Figure 5.49 Flat roof terminology

The amount of fall should be sufficient to clear water away to the outlet pipe(s) or guttering as quickly as possible across the whole roof surface. This may involve a single direction of fall or several directional changes of fall such as:

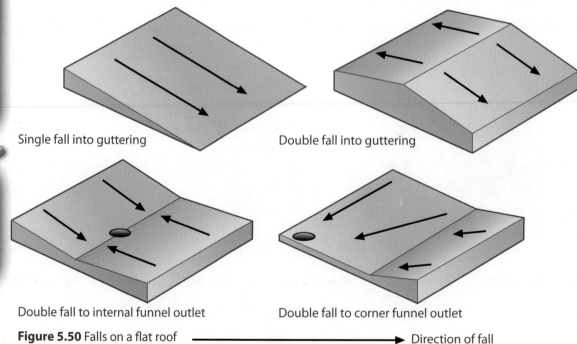

Single fall into guttering

Double fall into guttering

Double fall to internal funnel outlet

Double fall to corner funnel outlet

Figure 5.50 Falls on a flat roof ————————➤ Direction of fall

Pitched roofs

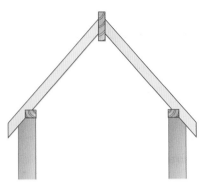

Figure 5.51 Single roof

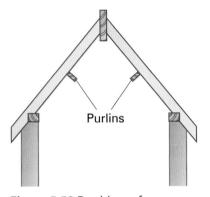

Figure 5.52 Double roof

There are several different types of **pitched roof** but most are constructed in one of two ways.

- **Trussed roof** – A prefabricated pitched roof specially manufactured prior to delivery on site, saving timber as well as making the process easier and quicker. Trussed roofs can also span greater distances without the need for support from intermediate walls.

- **Traditional roof** – A roof entirely constructed on site from loose timber sections using simple jointing methods.

Roof types

A pitched roof can be constructed either as a single roof, where the rafters do not require any intermediate support, or a double roof where the rafters are supported. Single roofs are used over a short span such as a garage; double roofs are used to span a longer distance such as a house or factory.

Safety tip

Roofing requires working at height so always use the appropriate access equipment (i.e. a scaffold or at least properly secured ladders, trestles or a temporary working platform).

Activity

Use sketches to show the difference between a single roof and a double roof.

There are many different types of pitched roof including:

- **mono pitch** with a single pitch
- **lean-to** with a single pitch, which butts up to an existing building
- **duo pitch** with **gable ends**

Figure 5.53 Mono pitch roof

Figure 5.54 Lean-to roof

Figure 5.55 Duo pitch roof with gable ends

Figure 5.56 Hipped roof

Figure 5.57 Over hip roof

Figure 5.58 Mansard roof

- **hipped roof** with hip ends incorporating crown, hip and jack rafters

- **over hip** with gable ends, hips and valleys incorporating valley and cripple rafters

- **mansard** with gable ends and two different pitches used mainly when the roof space is to be used as a room

- **gable hip** or **gambrel** – double-pitched roof with a small gable (gablet) at the ridge and the lower part a half-hip

- **jerkin-head** or **barn hip** – double-pitched roof hipped from the ridge part-way to the eaves, with the remainder gabled.

The type of roof used will be selected by the client and architect.

Trussed rafters

Most roofing on domestic dwellings now comprises factory-made trussed rafters. These are made of stress graded, **PAR** timber to a wide variety of designs, depending on requirements. All joints are butt jointed and held together with fixing plates, face fixed on either side. These plates are usually made of galvanised steel and either nailed or factory pressed. They may also be **gang-nailed** gusset plates made of 12 mm resin bonded plywood.

One of the main advantages of this type of roof is the clear span achieved, as there is no need for intermediate, load-bearing partition walls. Standard trusses are strong enough to resist the eventual load of the roofing materials. However, they are not able to withstand pressures applied by lateral bending. Hence, damage is most likely to occur during delivery, movement across site, site storage or lifting into position.

Wall plates are bedded as described above. Following this, the positions of the trusses can be marked at a maximum of 600 mm between centres along each wall plate. The sequence of operations then varies between gable and hipped roofs.

Definition

PAR – a term used for timber that has been 'planed all round'

Gang-nailed – galvanised plate with spikes used to secure butt joints

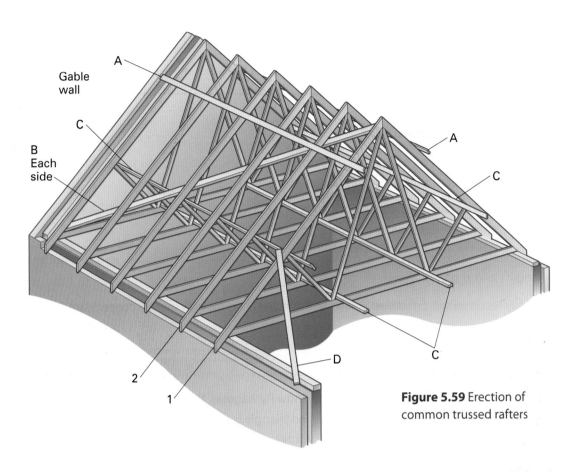

Figure 5.59 Erection of common trussed rafters

Step 1 Fix first truss using framing anchors 1.

Step 2 Stabilise and plumb first truss with temporary braces D.

Step 3 Fix temporary battens on each side of ridge A.

Step 4 Position next truss 2.

Step 5 Fix the wall plate and temporary battens A. Continue until last truss is positioned.

Step 6 Fix braces B.

Step 7 Fix braces C.

Step 8 Fix horizontal restraint straps at max 2.0 m centres across trusses on to the inner leaf of the gable walls.

Remember

Never alter a trussed rafter without the structural designer's approval.

Hipped roofs

In a fully hipped roof there are no gables and the eaves run around the perimeter, so there is no roof ladder or bargeboard.

Marking out for a hipped roof

All bevels or angles cut on a hipped roof are based on the right-angled triangle and the roof members can be set out using the following two methods:

- **Roofing ready reckoner** – a book that lists in table form all the angles and lengths of the various rafters for any span or rise of roof.

- **Geometry** – working with scale drawings and basic mathematic principles to give you the lengths and angles of all rafters.

The ready reckoner will be looked at later in this chapter, so for now we will concentrate on geometry.

Pythagoras' theorem

When setting out a hipped roof, you need to know Pythagoras' theorem. Pythagoras states that 'the square on the hypotenuse of a right-angled triangle is equal to the sum of the squares on the other two sides'. For the carpenter, the 'hypotenuse' is the rafter length, while the 'other two sides' are the run and the rise.

From Pythagoras' theorem, we get this calculation:

$A = \sqrt{B^2 + C^2}$ ($\sqrt{}$ means the square root and ^{2} means squared)

If we again look at our right-angled triangle we can break it down to:

A (the rafter length – the distance we want to know)

B^2 (the rise, multiplied by itself)

C^2 (the run, multiplied by itself).

Therefore, we have all we need to find out the length of our rafter (A):

$A = \sqrt{4^2 + 3^2}$

$A = \sqrt{4 \times 4 + 3 \times 3}$

$A = \sqrt{16 + 9}$

$A = \sqrt{25}$

$A = 5$

so our rafter would be 5 m long.

Did you know?

The three angles in a triangle always add up to 180 degrees.

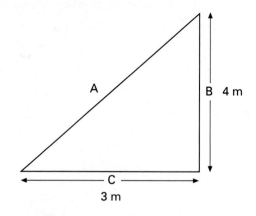

Finding true lengths

The next task is to find the hip rafter true length, plumb and seat cuts. This is done in two stages (the first will be familiar to you, as it is the same as for a common rafter).

The next step is to lengthen the common rafter true length by the amount of the rise, then join this line up to the base of the roof.

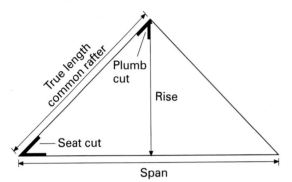

Figure 5.60 Finding common rafter true length

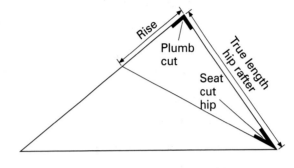

Figure 5.61 Finding hip rafter true length

Finishing a roof at the gable and eaves

There are two types of finish for a gable end:

- a flush finish, where the bargeboard is fixed directly onto the gable wall
- a roof ladder – a frame built to give an overhang and to which the bargeboard and soffit are fixed.

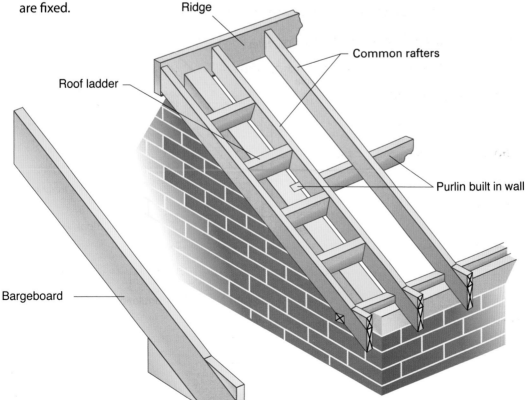

Activity

Use sketches to show how a gable ladder and bargeboard are fitted.

Figure 5.62 Roof ladder with bargeboard fitted

The most common way is to use a roof ladder which, when creating an overhang, stops rainwater running down the face of the gable wall.

The continuation of the fascia board around the verge of the roof is called the bargeboard. Usually the bargeboard is fixed to the roof ladder and has a built-up section at the bottom to encase the wall plate.

The simplest way of marking out the bargeboard is to temporarily fix it in place and use a level to mark the plumb and seat cut.

Eaves details

The eaves are how the lower part of the roof is finished where it meets the wall, and incorporates fascia and soffit. The fascia is the vertical board fixed to the ends of the rafters. It is used to close the eaves and allow fixing for rainwater pipes. The soffit is the horizontal board fixed to the bottom of the rafters and the wall. It is used to close the roof space to prevent birds or insects from nesting there, and usually incorporates ventilation to help prevent rot.

There are various ways of finishing a roof at the eaves; we will look at the four most common.

Flush eaves

Here the eaves are finished as close to the wall as possible. There is no soffit, but a small gap is left for ventilation.

Did you know?

Without soffit ventilation, air cannot flow through the roof space, which can cause problems such as dry rot.

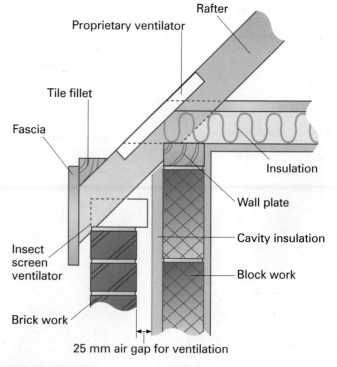

Figure 5.63 Flush eaves

Open eaves

An open eaves is where the bottom of the rafter feet are planed as they are exposed. The rafter feet project beyond the outer wall and eaves boards are fitted to the top of the rafters to hide the underside of the roof cladding. The rainwater pipes are fitted via brackets fixed to the rafter ends.

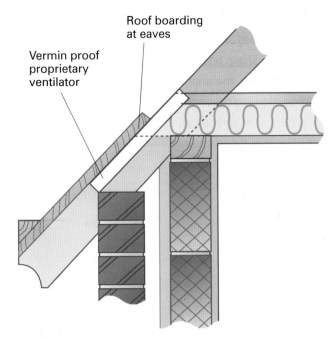

Figure 5.64 Open eaves

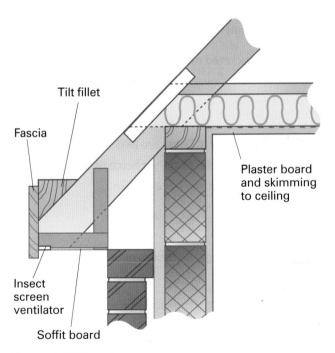

Figure 5.65 Closed eaves

Closed eaves

Closed eaves are completely closed or boxed in. The ends of the rafters are cut to allow the fascia and soffit to be fitted. The roof is ventilated either by ventilation strips incorporated into the soffit or by holes drilled into the soffit with insect-proof netting over them. If closed eaves are to be re-clad due to rot you must ensure that the ventilation areas are not covered up.

Sprocketed eaves

Sprocketed eaves are used where the roof has a sharp pitch. The **sprocket** reduces the pitch at the eaves, slowing down the flow of rainwater and stopping it overshooting the guttering. Sprockets can either be fixed to the top edge of the rafter or bolted onto the side.

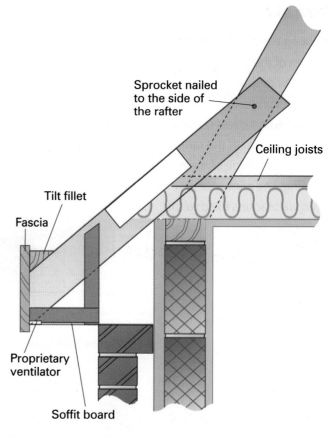

Figure 5.66 Sprocketed eaves

Roof coverings

Once all the rafters are on the roof, the final thing is to cover it. There are two main methods of covering a roof, each using different components. Factors affecting the choice of roof covering include what the local weather is like and what load the roof will have to take.

Method 1

This method is usually used in the north of the country where the roof may be expected to take additional weight from snow.

1. Clad the roof surface with a man-made board such as OSB or exterior grade plywood.

2. Cover the roof with roofing felt starting at the bottom and ensuring the felt is overlapped to stop water getting in.

3. Fit the felt battens (battens fixed vertically and placed to keep the felt down while allowing ventilation) and the tile battens (battens fixed horizontally and accurately spaced to allow the tiles to be fitted with the correct overlap).

4. Finally, fit the tiles and cement on the ridge.

Method 2

This is the most common way of covering a roof.

1. Fit felt directly onto the rafters.

2. Fit the tile battens at the correct spacing.

3. Fit the tiles and cement on the ridge.

Another way to cover a roof involves using slate instead of tiles. Slate-covered roofing is a specialised job as the slates often have to be cut to fit, so roofers usually carry this out.

Knowledge refresher

1 What is the definition of a flat roof?

2 What is the difference between a truss roof and a cut roof?

3 Explain the difference between a hip and a gable end.

4 What is the purpose of a roof ladder?

5 Describe an 'I' type joist.

What would you do?

You have been asked to quote for building a garage. The client is unsure about what type of roof to have. They ask your opinion. What type of roof would you select? Why would you select that type of roof? What could influence your selection?

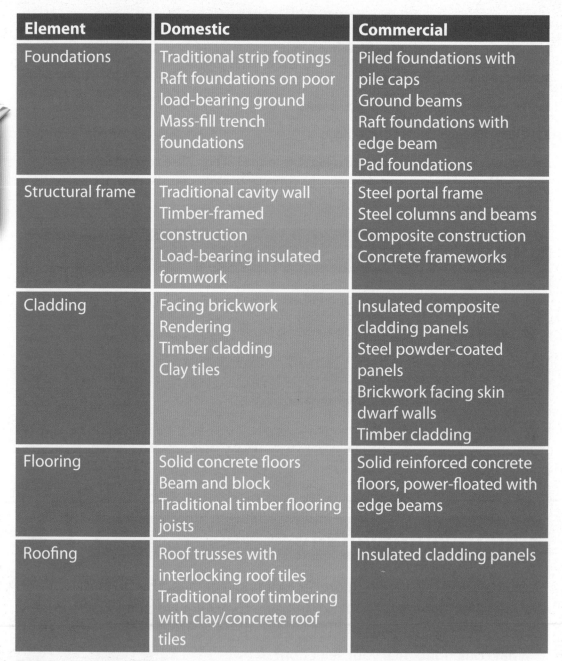

Find out

What is a steel portal frame used for in commercial building construction?

Element	Domestic	Commercial
Foundations	Traditional strip footings Raft foundations on poor load-bearing ground Mass-fill trench foundations	Piled foundations with pile caps Ground beams Raft foundations with edge beam Pad foundations
Structural frame	Traditional cavity wall Timber-framed construction Load-bearing insulated formwork	Steel portal frame Steel columns and beams Composite construction Concrete frameworks
Cladding	Facing brickwork Rendering Timber cladding Clay tiles	Insulated composite cladding panels Steel powder-coated panels Brickwork facing skin dwarf walls Timber cladding
Flooring	Solid concrete floors Beam and block Traditional timber flooring joists	Solid reinforced concrete floors, power-floated with edge beams
Roofing	Roof trusses with interlocking roof tiles Traditional roof timbering with clay/concrete roof tiles	Insulated cladding panels

Table 5.1 Comparison between domestic and commercial construction techniques

chapter 6

Energy efficiency and sustainability

OVERVIEW

Many of us are aware of the growing concerns around global warming and the current focus on minimising our 'carbon footprint'. This chapter deals with some of the factors that may be contributing to climate change, and the way in which modern construction methods and design can help to tackle it by creating a more sustainable environment – an environment that provides for the needs of the present without compromising the needs of future generations.

Energy efficiency and the reduction of wastage are now goals for every building project, and must be considered carefully, from the design stage right through the building process.

In this chapter, you will look at the modern materials and methods used to create structures in an energy-efficient way, with sustainability in mind.

This chapter will cover:

- the impact of climate change
- the role of the construction industry in tackling climate change
- sustainable construction methods
- environmental design considerations
- insulation
- how sustainability affects the different elements of a structure
- the energy efficiency and sustainability of different materials and components
- making materials last longer.

Climate change

Although much controversy surrounds climate change, it is hard to ignore the changes in weather patterns that we have witnessed recently, not just in the UK but also around the world. In the UK, these changes have resulted in extensive flooding, with winter months becoming warmer and summer months becoming wetter. Around the world, the ice caps are melting at worrying speeds, sea levels are rising and adverse weather systems have resulted in more frequent tornados and tropical storms.

The changing climate will certainly impact on the way we design buildings in order to cope with some of the possible effects, which include:

- rising temperatures
- rising sea levels
- rising amounts of rainfall
- higher humidity levels.

Area of land under flood

Although not fully proven, a number of factors are thought to contribute to global warming. The main causes are considered to be burning fossil fuels and the high levels of CO_2 being emitted into the air. CO_2 is emitted in many different ways including through car emissions, aerosol gases and burning untreated waste products.

Many practices have been introduced in an attempt to reduce the carbon footprint, and many more initiatives are being planned, including:

- recycling of packaging such as plastics, paper, cardboard, metals and glass
- use of alternative fuels to power cars and industrial machinery
- use of natural resources such as wind to supply electricity
- transport sharing to reduce the number of cars on the road, thus reducing CO_2 emissions.

Definition

Carbon footprint – total amount of CO_2 emissions produced by individuals and industry

New legislation has been introduced to make sure that the industry works towards the design of energy-efficient buildings, but at the same time the construction industry is becoming more conscious of the need to address climate change itself, by introducing its own initiatives and practices to minimise its carbon footprint.

Find out

Identify other initiatives and practices currently being used or developed in an attempt to reduce people's carbon footprint.

- The construction industry is putting greater emphasis on reducing building waste, and is being proactive in recycling many materials that would once have been disposed of to the detriment of the environment.

- There have been significant changes in the design of both domestic and industrial buildings.

- More natural resources are being used in the construction of buildings which, when the building is demolished, can be reused on a new development.

- Buildings now have improved insulation properties, thus reducing fuel usage for heating.

- There are alternative methods of providing energy for a building, including the use of solar panels to provide natural heat.

There are many, many ways of making buildings eco-friendly and we will look at some of these later in this chapter.

Sustainable construction methods

Modern construction methods have been developed in response to the new legislation on sustainability.

Cavity walling

Thirty years ago, it was quite acceptable to build a cavity 75 to 85 mm wide that had very little insulation included in its construction. Modern methods of cavity wall construction now allow for a cavity that is 100 mm wide and is fully filled with glass fibre cavity wall batt insulation, with an engineered lightweight internal skin of blockwork. In Chapter 5, you saw how spaced stainless steel wall ties are used to provide structural integrity to the completed wall, and how the insulation is not clipped.

In commercial buildings, cavity walls tend to be used in the bottom storey, while portal-framed construction with insulated cladding is usually used for the upper half of the unit.

Solar panels on a domestic dwelling

Timber-framed construction

Timber-framed construction is a fast, easy and efficient method of producing domestic buildings, and is sustainable in its approach as it uses timber from managed forests.

With timber-framed construction, the insulation is placed internally between the studs and covered with a DPM and plasterboard. A traditional skin of brickwork clads the exterior, and a cavity can be formed by covering the plywood panels with a breathable membrane. Stainless steel ties are screwed to the plywood to provide support for the brickwork.

The floors and roof are traditionally constructed, but the timber-framed panels can be prefabricated off site, which can have its own implications for energy efficiency.

Alternative construction methods

A number of other energy-efficient construction methods are growing in popularity.

Insulated concrete formwork

This is the 'Lego brick' construction method where hollow, moulded polystyrene forms are literally snapped together to form a wall, with the help of locating **castellations** on each form. Reinforcing rods are added where required and concrete is poured into the moulds and allowed to set. A solid wall is formed that can be rendered externally and plastered internally. Traditional brick cladding can be used externally to clad the structure.

Insulated concrete formwork is faster and more energy-efficient than more traditional construction methods as it uses less concrete, saves on site resources (for example, no bricklayers are needed), and the insulation is included as part of the formwork. Once constructed, the formwork can simply be rendered and plastered.

Energy-efficient insulated formwork

Insulated panel construction

This is a system that uses insulation bonded to the plywood panels that form a structural wall. Traditional cladding – both external and internal – can then be used to cover the insulated panel frames, using the same process as with the timber-framed construction.

This construction method uses pre-formed factory panels that can be readily assembled on site, saving time and resources, and finishing trades can come in earlier than usual, while the outer skin is being completed. The insulated panels are also highly thermally efficient.

Thin joint masonry

This is a system that uses high-quality dimensioned blocks that are up to three times the size of normal blocks. A cement-based adhesive is used to bond the locks together using 1 mm tight joints. The starting base course of brickwork has to be very accurate in its setting out and level. Traditional cavities are formed using helix type wall ties that are just driven into the thin joint masonry using a hammer.

This uses lightweight, thermally efficient blocks with no mortar, so it is faster and cheaper to construct, there is less waste, and the end product is very energy efficient.

Environmental design considerations

The design of a building or structure should now take on elements that make it environmentally friendly, both in its construction and in the way the occupants use it. The design should also take into account the needs of the local community, in terms of the infrastructure that will be needed to support its eventual construction and use.

Environmental considerations may include some or all of these elements.

- **The design brief** – This should aim to create a design that lowers pollutants to the atmosphere, reduces waste in its construction, reduces noise pollution, and gives the local community something that it can enjoy.

- **Recycling materials** – The design should specify the inclusion of recycled materials into the structure and aesthetics of the building project: for example, the use of crushed hardcore from demolition on site, or the reuse of slate roofing materials.

- **Energy efficiency** – The amount of **embedded** and used energy must be carefully considered for each element of the design, from boiler management systems to highly engineered aerated concrete thermal blocks.

- **Sustainability** – The design must contain elements that will meet the needs of future generations. A long, maintenance-free life span for the building is essential. Spending more at the initial cost stage can pay dividends in the future by reducing the cost of energy use and maintenance later on.

Activity

There are several major block manufacturers which produce the thin joint masonry system. Find one of their websites and produce a short presentation that includes:

- the advantages and benefits of the system
- why this system is sustainable
- a photograph showing how the blocks are joined.

Definition

Embedded energy – the amount of energy that has been used to create and manufacture the material and transport it to site for inclusion in the structure

- **Green materials** – The use of green materials is a vital environmentally friendly way of ensuring minimum impact on the local and global environment. For example, cedar timber boarding is a sustainable timber product that does not require chemical treatment or painting.

Knowledge refresher

1. Your client is considering moving away from traditional brick cavity walls to produce domestic dwellings. You have been asked to provide details of two alternative methods. Name two alternative modern construction methods that could be used.

2. Thin joint masonry does not use mortar to bond the blocks together. What product does it use?

3. What is the width of the cavity in timber-framed construction?

4. How is insulated concrete formwork finished externally?

5. What size is a modern cavity width in millimetres?

Insulation

The term 'building insulation' refers broadly to any object in a building used as insulation for any purposes – the term applies to both acoustic insulation and fire insulation as well as thermal insulation. Often an insulation material will be chosen for its ability to perform several of these functions at once.

Insulating buildings is vital: maintaining acceptable temperature in buildings (by heating or cooling) uses a large proportion of a building's total energy consumption, and the insulation can help to reduce this energy use.

The effectiveness of insulation is commonly related to its **R-value**. However, the R-value does not take into account the quality of construction or local environmental factors for each building. Construction quality issues include poor vapour barrier and problems with draught-proofing. Local environmental factors are simply about where the building is located. With cold climates, the main aim is to reduce heat flow from the building; with hot climates, the main aim is to reduce the heat from entering the building, usually through solar radiation, which can enter a building through the windows.

Did you know?

In a good environmental design, you can include a rainwater harvesting system that collects the rainwater from guttering and uses it to flush toilets and to wash down. This is known as 'grey water'.

Find out

What is a portal frame?

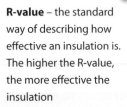

Definition

R-value – the standard way of describing how effective an insulation is. The higher the R-value, the more effective the insulation

Remember

The R-value is the inverse of the U-value.

Materials used in insulating homes

There are essentially two types of building insulation: bulk insulation and reflective insulation.

- **Bulk insulation** – this blocks conductive heat transfer and convective flow, either into or out of a building. The denser the material is, the better it will conduct heat. As air has such a low density, it is a very poor conductor – and therefore a good insulator. This is why air trapped between two materials is often used as an insulator, as in cavity walling.

- **Reflective insulation** – this works by creating a radiant heat barrier in conjunction with an air space, to reduce the heat transfer across the air space. Reflective insulation reflects heat rather than absorbing it or letting it pass through. It is often seen in the form of reflective pads placed behind radiators to reflect the heat from the rear of the radiator back into the room.

Forms of insulation

There are various forms of insulation, but the most common are:

- mineral wool/rock wool – these products are made from molten rock spun on a high-spinning machine at a temperature of about 1600°C, similar to the process of making candy floss, with the product being a mass of fine, intertwined fibres

- fibreglass – made in a similar way to rock wool, but using molten glass.

Mineral wool usually comes in sheets that are cut to fit between the rafters or studs. Fibreglass is similar, although it can come in rolls that can be cut to fit. Both materials are available in a variety of thicknesses to suit where they are going to be laid.

Other forms of insulation include:

- polystyrene sheets – polystyrene is a thermoplastic substance that comes in sheets which can be cut to size, again available in different thicknesses

- loose-fill insulation – this is particularly used where there is no insulation in the cavity walls: holes are drilled in the exterior wall and glass wool insulation is blown into the holes until the cavity is full.

Where to insulate

Where insulation should be used depends on the climate and the particular living needs, but generally insulation should be placed:

- in the roof space, between rafters or trusses and between joists

- on all exterior walls

- between the joists at every floor level, including ground floor

- in solid ground floor construction

- in partition walls

- around any ducts and pipes.

In the next section, you will see how insulation techniques can be applied to the different elements of a structure.

How sustainability affects the different elements of a structure

The sub-structure

Foundations need to be accurately set out to prevent wastage of materials: for example, the concrete in the foundations.

Cement manufacture is one of the processes that uses a high amount of energy and produces high carbon emissions. For example, by making a foundation larger than you require – perhaps because of an overdig by the excavator – you are using more material than you need to. As well as costing the client more, this will waste energy.

Sustainable excavation involves not removing the excavated material from the site. Removing excavated material incurs tip charges, which are part of the landfill tax. Dumping of waste into the land is now not a cheap option. By forming landscaped mounds on site, which can be planted and made into attractive green areas, you can increase the environmentally friendly impact of the excavation and save valuable resources in several ways, including reduced fuel charges, transport costs, air pollution and taxation.

Find out

Find out more about the landfill tax, on the internet or at your local library.

The depth to which foundations have to be taken is governed by Building Control Inspection

The depth to which foundations have to be taken is governed by Building Control inspection. However, optical and levelling equipment such as profile boards and a traveller must be used to make sure the foundation depth is set out accurately, which will reduce wastage and hence energy use.

Generally, the deeper the foundation, the more expensive and more energy-consuming it is to produce. Deep strip and wide strip are expensive, and it could be better to use piled/ground beam foundations and raft foundations. These foundation types do not require deep, wasteful excavations on site. The soil report and site conditions will need careful consideration at the design stage to produce an energy-efficient design that supports the building's loads safely.

What would you do?

Your client is adamant that all excavated materials are to be removed from site. There is a large, open space at the rear of the development. How would you convince the client of the benefits of leaving and landscaping the excavated material on site?

External walls

Cavity walls

Cavity walls have been used for the past 60 years. They rely on the external brick skin to keep the impact of the weather elements from crossing the cavity. The inner leaf of blockwork

Blockwork, brick and insulation used in cavity wall construction

is now thermally engineered to trap as much air into its structure as possible, making it lightweight and easy to handle. Openings in cavity walls must be fitted into the construction properly to maximise their energy efficiency.

Insulation in cavity walls

Cavity walls are insulated mainly to prevent heat loss and therefore save energy. The *Building Regulations* tell us how much insulation is required in various situations, and in most cases this would be stipulated in the specification for the relevant project to obtain planning permission from the local council.

Cavity insulation can be either Rockwall or polystyrene beads.

There are three main ways to insulate the cavity:

1. total or full fill

2. partial fill

3. injection (after construction)

Total or full fill

Figure 6.1 shows a section of a total fill cavity wall. The cavity is completely filled with insulation 'batts' as the work proceeds. The batts are 450 mm × 1200 mm, are made of mineral fibres, and placed between the horizontal wall ties.

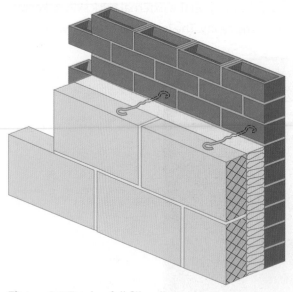

Figure 6.1 Total or full fill wall

Partial fill

Figure 6.2 shows a partial filled cavity, where the cavity insulation batts are positioned against the inner leaf and held in place by a plastic clip. More wall ties than usual are used to secure the insulation in place.

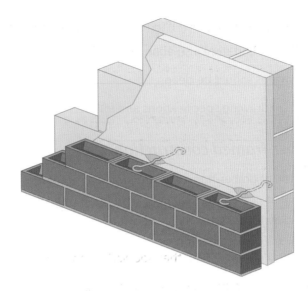

Figure 6.2 Wall with partial fill cavity

Injection

This is where the insulation is injected into the cavity after the main structure of the building is complete. Holes are drilled into the inner walls at about 1 m centres and the insulation is pumped into the cavity. The two main materials used are Rockwool fibreglass or polystyrene granules. The holes are then filled with mortar. If an older property were injected, then the holes would be drilled into the external mortar joints.

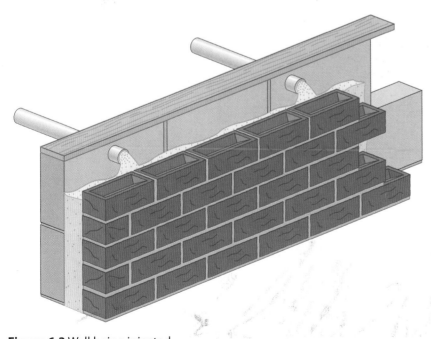

Figure 6.3 Wall being injected

Remember

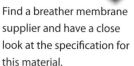

When injecting insulation, great care must be taken not to drill the bricks as they will be difficult and costly to replace.

Find out

Find a breather membrane supplier and have a close look at the specification for this material.

Remember

There must be a way for any trapped moisture to be 'breathed out', or the timbers could deteriorate.

There are three key points regarding insulating cavity walls:

1. Handle and store insulation material carefully to avoid damage or puncturing.

2. Cavities should be clean.

3. Read drawing specifications and follow manufacturers' instructions carefully.

Timber-framed construction

This type of timber-frame construction is a recent method of construction, and is highly efficient at preventing heat loss. In timber-framed construction, the traditional internal skin of block work has been replaced by timber-framed panels, which support the load of the structure. The breather membrane allows the passage of vapour in one direction so that the timber can 'breathe', but moisture is not allowed back in – the internal vapour barrier does not allow moisture to pass into the plasterboard.

The insulation within timber-framed construction is kept 'warm' by being placed within the warm side of the construction. It is protected by the vapour barrier and breather membrane.

The outside skin of a timber-framed building can be clad in several different materials, from a traditional brick skin of facing brickwork to rendered blockwork, which is painted.

Timber-framed construction

Solid walls

Occasionally, during refurbishment work, you may come upon an existing wall that is of solid construction. This tends to be in older houses from the 1890s, where solid brick walls were laid in English bond, with alternate courses of headers and stretchers. This method of construction did not contain a cavity or any insulation, and as such is not very thermally efficient.

Walls for modern-day uninhabited rooms – for example, an outbuilding – can be solid-block walls that are rendered externally to prevent water getting in.

Remember

Solid walls have to have high-quality brickwork and bonding with full mortar joints in order to prevent moisture getting in.

What would you do?

You are working on the first phase of a new housing complex. The construction type will be timber-framed, with a brick outer skin. You have started the first course above DPC and the general foreman has told you to place the insulation between the breather membrane and the brick skin. You are not happy about this. What might happen if you do what the foreman says? What should you do?

Material	Reasons for use
Bricks	Bricks are hardwearing, with a low porosity Available in a wide range of attractive colours and textures, which enhances aesthetics Can be used with coloured mortars to good effect Excellent lifespan
Blocks	Thermally efficient Lightweight Pre-textured ready to receive wet finishes High dimensional quality
Insulation	Essential to meet Part L of Building Regulations Sits within cavity – does not need clips if fully filled Thermally efficient component Relative low cost
Wall ties	Constructed of stainless steel so they do not corrode over time Resist lateral forces on the wall Increase the width of the cavity wall making it more stable
Timber frames	Lightweight construction method Very fast to construct Thermally efficient Sustainable construction material
Sheathing plywood	Used to provide strength to the timber frames External quality to resist moisture Strong material in shear Suitable to fix brick ties to
Breather membrane	Enables the passage of moisture Acts as a weather protector during construction Can be stapled to plywood easily
Vapour barrier	Prevents the passage of moisture, keeping the internal plasterboard dry Can be stapled to timber studs easily Retains the insulation in place
Plasterboard	Protects the DPM underneath Final finish to timber stud construction Fire-resistant material

Remember

Cost is just one of the factors that will affect the choice of a material.

Table 6.1 Reasons for using different materials in external walling

Comparison of insulation properties between cavity and timber-framed constructions

Existing cavity wall

A house that is over 50 years old will have been built with the level of insulation that was required at that time under legislation – which may mean none at all within the cavity wall. The insulation properties of the existing wall can be upgraded in three ways.

- The first is to use a foaming insulation, which is pumped in from the outside. You must take care with any gases that the insulation may give off.

- The second method is to use a blown glass fibre insulation, which is 'blown' under pressure through holes formed in the external brickwork to fill the cavity.

- The final method is to clad the internal walls with either insulation-backed plasterboard attached by dot and dab or timber battens clad in plasterboard. This method does reduce the internal dimensions of the room.

New cavity construction

This construction is much more thermally efficient, as you can look at reducing heat loss through the structure as a whole. Highly efficient internal blocks are used to form the internal skin, with a 100 mm cavity, which is fully filled using a mineral wool insulation material. The traditional brickwork skin is the same as for the older type of construction – this has not changed over the years.

Modern sheet insulation materials can also be used within cavity construction. Manufactured from high-performance rigid urethane, these materials use space age technology to resist the passage of heat, but may require clipping within a cavity to the wall ties. Therefore, the cavity is said to be partially filled, rather than fully filled.

Timber-framed construction

This is the most thermally efficient system. It uses insulation (usually quilted) fixed within the timber studs.

Again, rigid insulation products can be used to fill between the timber studs. You can increase the level of insulation by making the timber studs deeper in a mineral wall, or doubling up on rigid insulation boards.

Rigid insulation is normally covered with a foil surface that resists the passage of vapour and slows down the resistance of heat, reflecting it back towards the warm side of the construction.

Remember

Any insulation that is retro-fitted is normally put in by a specialist contractor after a building survey.

Find out

Find a rigid insulation manufacturer's details for a wall insulation board, and have a look at the fixing details.

Activity

To give a visual comparison between cavity and timber-framed construction, produce a sketch drawing of a cross-section through a wall, clearly illustrating the position of the insulation within the construction. Do this for either rigid or quilt insulation materials.

Internal walls

Internal walls can either be **load-bearing** or non load-bearing. If they are load-bearing, they must be made of materials that can resist the load. In this section, you will investigate the sustainability and energy efficiency of the internal walls and linings most often used in modern construction.

Fair-faced blockwork

This is economical in terms of the energy used to produce the desired finish: the blockwork produces the completed finish itself, and no secondary wet or dry finishes (plastering or plasterboard) are applied. If the wall is not to be painted, a good quality block must be chosen; if it is to be painted, paint-grade blocks are needed. You must consider the joints between blocks carefully when looking at the type of pointing required. This must be neat and applied to both sides of the wall.

Fair-faced concrete blockwork is an ideal solution for heavily trafficked areas, sports halls and other walls that need a hardwearing surface. However, it may take several coats of paint to produce an acceptable finish on the blockwork surfaces.

Timber stud walls

A traditional timber stud wall construction

These are constructed using regulated timber studs of equal widths. These are attached to head and sole plates, at the top and floor level respectively. Noggins are added to aid stability and fixing of finishes.

Timber studs are a sustainable timber product. Putting up a timber stud wall is an efficient process, as the walls do not have to dry out and no scaffolding is needed. The studs are normally clad with a plasterboard skin on each side, which can be dry-lined or skim-plastered. With this type of wall, it is also quick and easy to distribute services around a building.

Metal stud walls

In this system, timber studs are replaced by preformed metal channels. These are fixed together by crimping (using a special tool) or screwing. Metal studs are made with recycled steel, but do have to go through a galvanising process. However, their strength-to-weight ratio is better than timber. Plasterboard finishes are normally screwed into the studs using dry wall screws and a hand-held drill.

Both timber studs and steel channels can be recycled after final use.

What would you do?

The architect is unsure which of the three systems to incorporate into the ground floor internal layout design. You have been asked to discuss the fixing side of the three methods. Prepare a few notes so that you can discuss these issues with your tutor, who will act as the architect.

Dry lining

This is the process that lines a surface – normally blockwork – with a plasterboard finish.

'Dot and dab' is a dry lining process. Plasterboard adhesive is applied in dabs to the surface of the wall, and plasterboard is pressed onto the adhesive dabs, then knocked until vertical and aligned with the wall, using a straight edge. A temporary fixing is often used to hold the board in place, and then removed once set. The joints of the plasterboard are bevelled so that a jointing finish tape and filler can be applied to complete the wall.

This process leaves an air gap behind the board, which makes the wall more energy-efficient in terms of heat loss. This dry lining process is useful in refurbishment work, where untidy walls can be easily covered up and upgraded to a quality finish, saving the need to replace whole walls.

Plastered blockwork

This is the least energy-efficient process because the finish has to dry out, using dehumidifiers and a heat source if necessary (for example, in winter). Wet plaster cannot be painted. A plaster coat is normally applied in two layers, the second being a trowelled and polished final finish that produces a smooth wall. There are various types of plasters to suit the surface the plaster is being applied to and the level of wear the wall will be subjected to.

Material	Reasons for use
Fair-faced blockwork	• Economical single process • Can be left as a self-finish • High-quality block available • Saves on secondary resources and energy • Greater sound resistance
Timber stud walls	• No wet construction • No drying-out time • Renewable resource • Lightweight construction • Traps air within void that can be insulated • Easy hiding of services
Metal stud walls	• Can incorporate recycled steel into manufacture • Good weight-to-strength ratio • Screwed physical fixing of linings • Internal void can be insulated
Dry lining	• Covers up untidy backgrounds • Forms air void behind • Can be used with insulation bonded to plasterboard • Quick, easy method of providing a smooth finish • Does not use wet trades • No drying out period
Plastered blockwork	• Traditional finish • Hardwearing • Easy to repair

Table 6.2 Reasons for using energy-saving materials in internal wall construction

Flooring

Solid floors

Concrete floors are constructed like this.

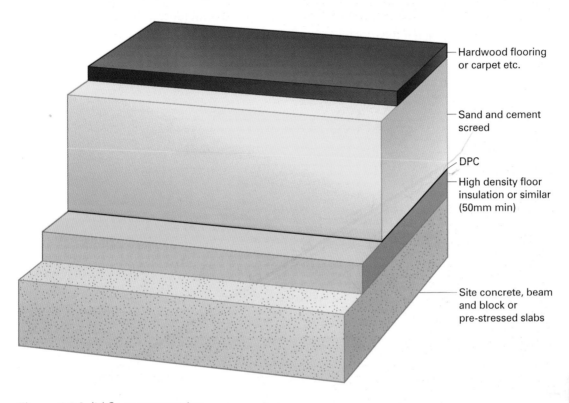

Hardwood flooring or carpet etc.

Sand and cement screed

DPC

High density floor insulation or similar (50mm min)

Site concrete, beam and block or pre-stressed slabs

Figure 6.4 Solid floor construction

The insulation can be placed under or over the concrete, and must be of the right specification to take the loadings from the floor.

Hollow floors

Beam and block floors require specialist thermal insulation that clips under the beams, below the level of the blocks.

Roofing

A roof serves several functions: it protects the building from the elements, directs water into gutters and makes the building look aesthetically pleasing. A roof must also meet the requirements of Part L of the Building Regulations, which detail the levels of insulation that must be included to achieve the required **U-value.**

Waterproofing, ventilation and insulation are crucial in making a roof efficient and long lasting, and contribute to the energy efficiency and sustainability of the whole structure.

Remember

The insulation manufacturer will be able to help you with typical installation details for placing the thermal insulation into a design.

Definition

U-value – a measure of thermal transmittance through a building component, usually a roof, wall, window, door or floor

Pitched roofs

You will remember from Chapter 5 that a pitched roof is made up as follows:

- Softwood rafters span from the ridge to the eaves and are fixed to a wall plate, which is strapped to the internal skin of the wall.

- The rafters are then covered with a felt or a breathable membrane, which stops moisture entering the building if the roof tiles are weakened.

- The felt or membrane is held in place with tile battens manufactured from treated softwood, which are fixed to the rafters using galvanised nails.

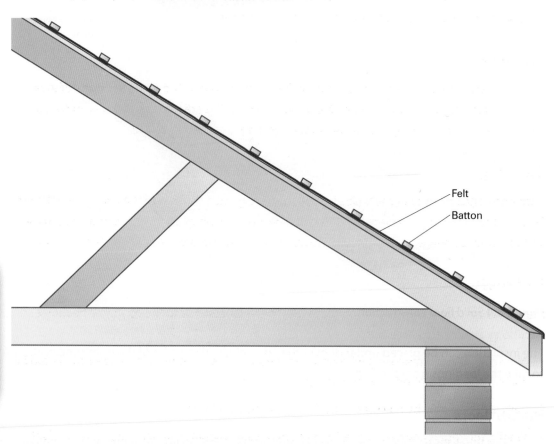

Felt

Batton

Figure 6.5 Cross-section of a pitched roof showing positioning of felt and battens.

Insulation

The insulation can be placed in two locations:

- between and over ceiling joists – up to 300 mm of mineral wool insulation (manufactured from glass or rock) is laid, 150 mm between the joists and 150 mm at 90° to this, to form the full thickness

- between or over rafters – form insulation boards are cut and fitted between roof timbers or, with a loft conversion, are laid over the joists before the tile battens and tiling are fitted.

Waterproofing, damp proofing and ventilation

The ridge is where the two rafters meet at the top of the roof. A fascia and soffit complete the eaves detail of the roof and make it waterproof. Ventilation is provided to prevent the air stagnating, which could lead to the roof timbers rotting. Ventilation may be provided to the roof by placing vents into the soffit or on top of the fascia. In this case, insulation is placed in two layers: one between the ceiling joists, and the other perpendicular to the first layer.

Valleys are formed where two roofs meet within an internal corner. Here a valley board is formed using plywood, which is covered with a flashing to stop water entering. All roof abutments, such as chimneys, must have flashings placed around them to make sure that water cannot get in, causing problems with damp and rot.

Roof tiles

Roof tiles tend to be chosen to fit in with the local environment, and are often made of local materials: for example, slate in Wales. Most modern roof tiles are manufactured from concrete and have an interlocking system to stop water getting into the roof space.

Guttering

With both pitched and flat roofs (see below), guttering must be fixed to the fascia board laid to falls, so that rainwater can be directed to fall pipes. The fall pipes connect the guttering to the surface water drainage system. Modern guttering is manufactured from uPVC or aluminum.

Did you know?

In the past, guttering used to be manufactured from timber or cast iron.

Flat roofs

Flat roofs have notoriously short life spans due to the low pitch and the continual heating and cooling cycle of the environment.

- Normally, felt manufactured with bitumen is used to cover a plywood decking and is laid in three layers, each bonded with bitumen.

- Falls are produced by nailing timber firrings to the roof joists.

- Finally, the roof is covered with a layer of white spar chippings, which reflect the sunlight and prevent solar heat gain from damaging the roof.

With flat roofs, the insulation is set between the joists within the ceiling void or as cut-to-falls insulation, which is laid over the plywood decking and covered with roofing felt. Both methods need a vapour barrier to be fitted, to resist the passage of moisture.

Asphalt is a naturally occurring product. When it is heated up to melting point it can be laid in a single layer onto the prepared roof, forming a single, impenetrable barrier.

Other elements of the structure

Insulating timber partitions/floors/roofs

Insulating between timbers – whether joists or studs – is simply a matter of placing either glass/rock wool or polystyrene between the timbers, with thicker insulation placed between joist roof spaces.

Figure 6.6 Insulating pipework using mineral wool.

Insulation being fitted between timber

Insulating water tanks and pipes

All water pipes in a loft space or on exterior walls must be insulated to protect them from freezing, which may cause them to crack. There are two methods used when insulating pipes: mineral wool matting or pre-formed moulded insulation.

- **Mineral wool mat** – a small mat is wrapped around the pipes with a bandage and secured with tape or string. The pipes must be completely covered with no gaps, and taps and stopcocks must also be covered.

- **Pre-formed moulded insulation** – this is available to suit different sizes of pipe, and specially formed sections are available for taps and stopcocks. The mouldings can be cut at any angle to fit around bends and when installed the sections of insulation should be taped together to ensure that they are fully enclosed around the pipes and are tightly butted up to one another.

Insulating water tanks

Any water tanks in the loft space should be insulated around the sides and on the top. The insulation around the sides must extend down to the insulation on the floor of the loft. Insulation jackets are available to fit most sizes and styles of tank.

Figure 6.7 Use of preformed mouldings to insulate pipework.

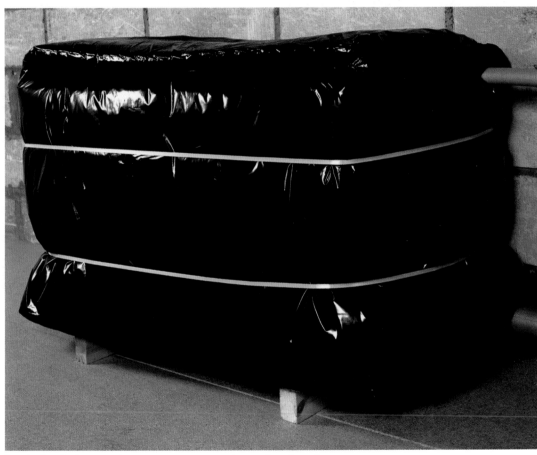

Insulation jacket used to protect water tanks against freezing

Preventing heat loss in windows and doors

A specialist **FENSA** installer should be used to supply and fit new uPVC windows and doors. These will meet the agreed insulation standard set by the Building Regulations. The cavity should be closed using a thermal bridging product that insulates across the cavity and acts as a DPM, if required.

Definition

FENSA – Fenestration Self-Assessment scheme, under which window installers agree to install to certain standards

Activity

Find out the thermal value of a same-sized piece of polystyrene and mineral wool then compare.

Knowledge refresher

1   How is the effectiveness of insulation measured?

2   What is the main reason for using insulation?

3   Describe the two main types of insulation.

4   How is mineral wool formed?

5   List four areas that require insulation.

6   Why should pipes be insulated?

What would you do?

You have been tasked with insulating a loft space. What type of insulation would you use? Why would you use that type? (Consider cost, workability, R-rating, etc.)

The energy efficiency and sustainability of different materials and components

How sustainable are the most common building materials?

As you work in the construction industry, you will come across a vast range of materials. But which of these are the most sustainable? The table below gives a quick overview of the most common building materials, looking at how they are made and any issues to do with their use and disposal.

Remember

Sustainability means meeting the needs of future generations, without depleting **finite** resources.

Definition

Finite – a resource that can never be replaced once used

Materials	Sustainability
Polystyrene	• Petroleum-based product • Can be recycled • Pollution caused through manufacture • Often used as packaging for equipment • High insulation properties
Polyurethane	• Made from volatile organic compounds that damage the environment • Often turned into foams • Gives off harmful chemicals if burned • Disposal issues: if the foam catches fire, it causes pollution and produces carcinogenic gases
Softwood	• Can be recycled into other products • A managed resource that can be grown over and over again • Provides a pollution filter, takes in carbon dioxide gives out oxygen • A natural 'green' product • Requires treatment to prevent rot
Hardwood	• Forms part of the tropical rainforest • Intensively felled • Takes a longer time to re-grow than softwood • Expensive resource
Concrete	• Cement production causes CO_2 emissions • Disposal issues – can only be crushed and used as hardcore • Hardwearing with a long lifespan • Relies on a lot of formwork and falsework
Common brick	• Manufacture involves clay extraction, which has environmental issues • Waste areas form ponds • Uses finite gas energy to fire bricks

Table 6.3 Sustainability of common materials

Materials	Sustainability
Facing brick	• Secondary process to produce the patterns and colours • Useful life of up to 100 years • Uses natural products
Engineering brick	• Hardwearing, with a long lifespan • Water resistant • Heavy to transport to site, so fuel costs increase
Aggregates	• Extracted from gravel pits, causing environmental damage • After extraction, water fills pits to produce another habitat
Glass fibre quilt	• Can use recycled glass in manufacture • High insulation value and long lifespan • High thermal insulation-to-weight ratio • Economical transport from factory to site
Mineral wool	• Manufactured from glass or rock • Provides a lightweight, high insulation product • Can be used in walls and roof spaces • Can use recycled glass
Plasterboard and plaster	• Disposal issues, creating landfill with high sulphate content • High wastage with cutting to size • Manufacture uses gypsum, a waste by-product • Recycling skip agreements in place with manufacturers
Concrete block	• Uses cement • Can incorporate waste products from power stations • Sustainable manufacturing process • Take-back recycling schemes for waste blocks
Thermal block	• High thermal efficiency • Low weight-to-strength ratio • Uses cement in manufacture additives to produce a block full of air bubbles
Metals	• Can use recycled scrap steel in manufacture • Requires secondary treatment to prevent rust • Non-ferrous metals have a long lifespan
Glass	• Can be completely recycled • Waste from cutting can be recycled • Issues with recycling: toughened glass has been heat treated as a secondary process, so needs special recycling techniques

Table 6.3 (cont.) Sustainability of common materials

Using sustainable materials

The following materials are the most sustainable of those listed in Table 6.3. You will now take a closer look at their use in domestic and commercial dwellings.

Softwood

This naturally occurring timber product is grown more in Scandinavian countries, where growth is slower, producing a more structured wood grain. It must be processed then transported to the UK for further manufacture into timber products. Softwood should be purchased locally to save on transport costs. Timber can be used in the following areas of a building: the roof, floor joists and flooring, second fix joinery products, stud walls, windows and doors. Timber can be completely recycled into a further product, such as chipboard.

Using softwood to clad the outside of this building is an attractive and sustainable method that requires no treatment

Concrete

This is only sustainable in that the finished product has a very long life. Cement production is costly to the environment in terms of energy used and emissions released to the atmosphere. Concrete can be recycled into a crushed hardcore but no recycled materials can be used in its manufacture. Concrete is used in foundations, lintels, floors and some roof structures.

Common bricks

Common bricks are of lower quality and are used in areas where strength is required but facing bricks aren't needed. They would be used as coursing bricks in internal skins of external walls and as levelling courses under floor joists.

Facing bricks

These are used where you can see the external brickwork as they look more attractive, but they are more expensive to make. They are mainly used for the external skin of a cavity wall or for garden walls where these have to match the house.

Engineering bricks

These are hardwearing high-strength bricks that are water-resistant. They are used below DPC level as they are not affected by rising damp. They are an excellent brick to use within drainage manhole construction.

Activity

Analyse different types of materials as listed in Table 6.3 used in the construction of domestic and commercial dwellings

Find out

Find a local ready-mixed concrete supplier and establish what concrete products they can supply.

Find out

Visit a brick manufacturer's website and find out about their sustainable practices.

Aggregates

These are used for several purposes. Primarily they are constituents of concrete mixes, where a blend of coarse and fine aggregate plus cement and water is mixed to form concrete. A more modern application of aggregates is in external landscaping, where they are used to provide attractive areas that resist water run-off into the main drainage system, thereby adding to sustainable drainage for surface water.

Glass fibre quilt and mineral wool

These are used to insulate houses and commercial properties. They are used within ceiling voids above and between the joists. External walls use pre-cut 'batts' of mineral wool, which are built at full thickness into the cavity. Internal walls can also use mineral wool, but it must be of a heavier density, so as to resist the passage of sound. This type of insulation is very thermally efficient and will reduce heat loss from a building.

Example of glass-fibre quilt

Did you know?

You can use sheep's wool as a modern insulation product. It is a fantastic insulator and can 'breathe out' moisture.

Safety tip

Always use a dust mask and gloves when handling fibre insulation products.

Metals

Using stainless steel in many building materials (such as wall ties) helps overcome corrosion problems. Mild-steel products require a secondary process, which could be galvanising, powder-coating or just a layer of paint. Metal is used in other products such as lintels, joist hangers and roof truss fixings, and in aluminum windows. Non-ferrous metals are used in water pipes and fittings.

Glass

Glass is a versatile material, used mainly in windows. In certain areas, only safety glass can be used. Glass can be manufactured from up to 80% recycled product. It is transported in standard sheet from the factory to the supplier, who then cuts it to size locally. Any waste can be fully recycled, either locally or via the manufacturer.

Selecting appropriate materials for sustainability

When choosing materials, you should consider a number of factors that impact on energy efficiency and sustainability:

- **Is it recyclable?** Ideally the material selected should be recyclable, at least at the end of its life, as should any wastage generated during its use or installation.

- **Is it local?** If the material is available locally, this reduces the cost of transporting it from the supplier to the construction site.

- **What are the transport implications?** The weight of the product will have an effect on the cost of its transport and the amount of fuels expended in getting it to the site.

- **How much waste is there?** A product that produces no wastage is valuable in terms of not having to throw away valuable resources and the energy that they have used in manufacture.

- **Is it a natural product?** A product that is produced naturally, for example, growing timber can be replenished time and time again.

- **Is it environmentally friendly?** – A material that does not damage the environment in its manufacture and use is a sustainable product that is often referred to as a 'green product'.

- **What is its lifespan?** A longer lifespan, with as little maintenance as possible, means that less energy and resources will have to be put into the product in the future, which saves valuable future resources.

- **Is it worth spending more now?** A material that does not need, for example, painting every five years will save on energy time and costs. Spending a bit more money now will save future expenditure.

What would you do?

You are working for a client who builds doctors' surgery centres. The current contract you are working on is a large medical centre. You have been asked by the client to recommend the finishes for the front of the centre due to your experience in using current sustainable materials. Assist the client by discussing the advantages of cedar boarding over rendered blockwork.

Making materials last longer

Many of the materials used in a building project can have their lifespan reduced by the effects of damp, water, frost and chemicals. Treating problems like this early – or, better still, preventing them from happening in the first place – is part and parcel of good environmentally aware building practice.

The effects of the elements on various building materials

Masonry

Efflorescence is one cause of the effects of moisture migrating within a new brick structure. Water moves through the mortar into the brick work causing salts to move to the outside of the newly constructed wall, this process is known as primary efflorescence. The water evaporates leaving a white deposit on the surface of the brick. It can be brushed off and the action of weathering will eventually remove the effect.

An example of efflorescence on brickwork

Concrete

The effect of water on concrete is to discolour its initial fresh new look from the airborne pollution that is present in the atmosphere and which is picked up within rainfall. Water staining on concrete where the designer has not detailed for the run-off from the building is another obvious effect. Water action within less porous concrete can result in the leaching of alkaline ('lime leaching') through the concrete that form deposits very much like those within cave structures.

Timber – wet and dry rot

This is caused by two different elements. Wet rot is caused by excessive exposure to moisture, which eventually causes the breakdown of the timber cells so that they rot and deteriorate. Dry rot is caused by a fungus that sends out long threads, which attack and eat away at the cells of the timber, causing structural damage.

Find out

Have a look at some of the newly constructed houses in your location and see if you can spot some efflorescence appearing on new brickwork.

Remember

All of the above defects can be remedied through good design, quality workmanship and materials and effective maintenance.

Metal

Rust is caused by the oxidation of a ferrous metal. For the reaction to occur, three things must be present: metal, water and oxygen. Rust forms as an orange deposit on the surface of the metal. If it goes untreated, it will continue to eat away at the metal, eventually compromising its strength.

Particular problems caused by the elements

Spalling

Spalling is the action of freezing water on the surface of a porous brick. The freezing action causes water to expand, pushing flakes of brickwork from the brick face. The photograph clearly illustrates this action, which leaves the stronger mortar joints proud of the original brickwork face.

The effects of spalling on brickwork

Thermal expansion

The Building Regulations state clearly the specification for expansion joints within brickwork external walls. These allow the brickwork to expand in the summer months and contract in the winter months without showing signs of cracking. Thermal movement may appear as complete cracks across bricks vertically.

Timber and concrete are good at resisting the affects of frost or freezing water. Exposed timber will maintain its structural integrity for many years. Concrete, once damaged, allows water to enter; if this meets any exposed reinforcing bars, then problems will occur.

Damp

Damp can cause many problems in a building, and the damage it causes can be costly and difficult to repair. A range of methods and materials can be used to help stop damp getting into a structure.

DPCs – placed within the skins of the external walls; always 150 mm above finished ground level.

Slate – used traditionally as a barrier to resist the passage of moisture; still found in older external cavity walls; can still be used as a DPC, but commonly replaced by plastic DPC.

Engineering brick – Grade A quality brick, so dense it naturally resists the passage of water; can be used below DPC to act as an additional barrier to rising damp.

Pitch polymer – a bitumen-type DPC sometimes seen in external walls; can squeeze out of the joint due to pressure and heat; seldom used today.

PVC – the most common type of DPC in use today; economical and easy to bed into the joint; often textured to bond to the mortar above and below the joint.

Did you know?

Brickwork expansion joints will need taking right through the cavity and the internal block skins.

Find out

What do the terms 'Grade A' and 'Grade B' mean on engineering bricks?

Safety tip

With any chemical process, always refer to the supplier's safety data sheet for instructions.

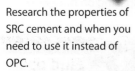

Injected – chemical DPC is injected into the wall to prevent the passage of moisture; used as a cheap refurbishment technique as you do not have to remove brickwork; internal plaster often removed up to 1 m high and wall replastered with a renovating plaster after injection work is complete.

DPMs – placed under floors to prevent rising damp from entering warm floor; tucked into the DPC within the external wall, thus 'tanking' the construction against damp.

Visqueen – common trade name for a 1200-gauge plastic product tough enough to withstand tearing but which will bend around corners. Note: hardcore beneath DPM should always be sand-blinded to prevent rupture of DPM.

Bituthene – a sheet material manufactured from bitumen that can be laid in sheets, lapped, and used both horizontally and vertically; mainly used for waterproofing basement walls; must be used in accordance with manufacturer's instructions.

The effects of chemicals on building materials

Concrete can be attacked by certain chemicals.

Carbon dioxide from the air can react with certain chemicals in the concrete. This decreases the protective alkaline which ensures that the steel reinforcement does not rust. Once this bond is broken, then rusting can occur as the concrete expands and spalls away from the reinforcement bars.

Sulphates that come into contact with the cement in concrete can attack the chemical bond of the cement, affecting its strength. **Sulphate-resistant cement (SRC)** and not **ordinary Portland Cement (OPC)** must be used where this is a risk: for example, where groundwater is contaminated.

Sulphates can also attack the brickwork cement mortar, especially with a chimney where coal is being burned. Sulphates eat away the cement mortar joints, the reaction causes the joint to expand and the chimney can start to lean over.

Chlorides have been used in the past to alter the setting times of concrete. Unfortunately, over time these affect the strength of concrete, which can have a disastrous effect if the material is under excessive loading conditions.

Alkali silicate is present in some concrete aggregates. This can react with the cement and water and expand, causing a pop-out of the concrete surrounding the reaction. This can have an adverse effect on structural concrete.

Acid rain – a product of burning fossil fuels – also has an effect on certain building materials. Natural stone is mostly affected, as acid rain eats away at the surface of the stone, causing long-term damage.

Effects of heat and fire on building materials

Masonry

Brickwork and blockwork cope well in a fire within a dwelling or commercial property. Extreme heat can cause masonry to expand and crack but this would be after a substantial time from evacuation of a building. Fire is used in a kiln to harden bricks so they are more than capable of staying stable in a fire. Smoke does, however, cause blackening to the surfaces of bricks and blocks. This can to some degree be washed off but may leave a permanent stain that requires decoration internally and power washing externally.

Water, as we have seen previously, when combined with a frost or freezing conditions causes spalling to the surface of the brickwork or the breakdown of the mortar joint due to weathering.

Concrete

Concrete can spall under the influence of fire. If the reinforcement lies near to the surface of the concrete and this heats up then it will expand at a different rate to the surrounding concrete. This expansion can cause cracking to the concrete structure. Concrete is fairly resilient to water damage and will dry out after wetting; but if the water contains contaminants then this can lead to the chemical attack of the cement within the concrete.

Timber

Heat and the presence of oxygen cause the combustion of timber by fire. The surface chars and eventually breaks down the structural integrity of the timber until it is burnt right through. Smoke damage can discolour timber which will then require decoration if it has not caught fire and charred. Water can damage timber by wetting which expands the hygroscopic material and causes dimensional change to the timber, which will eventually rot if this wetting persists.

Metal

Metal does not react well in a fire. As it heats up the molecular structure weakens and it loses up to half its strength at over 500 degrees. This can cause the collapse of structures as the steel melts slightly and warps under extreme heat. However, this does take some time and may not affect the evacuation of a structure. Water reacts with exposed metal, as we have seen, to form rust. Surface rust is not harmful but continual exposure to the elements of unprotected metal will result in severe corrosion.

Find out

Consult your local newspaper's website and look up a building fire that has occurred in your area. Look for photographs to see what damage was caused.

Remember

Building Regulations will make provisions for the building structure in the event of a fire.

Treating building materials to prevent the effects of deterioration

Intumescent paint

This is a chemical-based paint that reacts with high temperatures. It foams and expands around the steelwork it is painted upon to protect it from the heat giving the occupants time to escape.

Treated timber decking

Paints

These are used to protect timber, brickwork, and metal work from the effects of water. They form a microscopic seal across the surface preventing water penetration past this layer.

Water repellant

This can be painted onto the surface of exposed brickwork to allow water to run off the brick face more easily. This prevents it standing and absorbing into the brickwork. Where if it freezes, it can become a problem as we have seen.

Vac Vac and tanalising

These are chemical pressure impregnation systems to chemically treat timber from fungi and insect attack. They fill the pore holes within the timber and kill off any wood-boring insect that attacks its structure. Fungi includes the dry rot spores and other cellulous eating organisms.

This timber treatment alters the colour of the natural timber often to a green shade.

Sulphate-resistant cement (SRC)

This is a cement that contains chemicals and mineral aggregates which resists the attack of sulphates within concrete and is used where this is detected by testing.

Injected damp proof courses

These use chemicals of various kinds and mixtures to form a chemical barrier within the brickwork after they have cured and filled the pores in the bricks. These have to be injected through holes drilled into the outer wall and sometimes in inner walls.

Rectifying and preventing deterioration

Masonry and concrete

Masonry and concrete make a movement joint that allows the brick and block panels to expand and contract under thermal movement. A compressible board is used within the joint which is sealed with a polysulphide joint to prevent water ingress. Joints within concrete slabs are very similar but the Flexcell is replaced with a material that is more resilient.

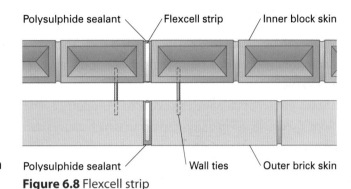

Figure 6.8 Flexcell strip

Timber

Cutting and splicing in new timber is an age-old prevention method that saves having to remove the whole length of timber, manufacture a new piece and refit. Timber can be initially protected by pressure impregnation or it can be simply painted or stained using a flexible coating as the timber will expand and contract. This method of coating must be maintained to extend the lifespan of the timber and any damaged areas repaired quickly.

Insecticides or fungicidal washes will remove mould growth on timber but ventilation and quick drying of wet surfaces is the key to preventing this damage.

Methods used on metals to give galvanic protection and protective coatings

Galvanising using a hot dip process provides a good layer of protection to ordinary mild steel. The steel is fed through a bath of molten zinc, which coats the metal and gives it long lasting protection against water damage.

High-quality paints that are developed for bonding to steel work should be used instead of galvanising where colour is required. A primer base coat is applied followed by several other coats that build up the film of paint. Powder coating steelwork is another process that can be used to coat the steel with a colour protective layer.

Find out

Have a look around at the built environment where you live and work. Analyse the brick and concrete buildings are see where the movement joints have been formed.

Find out

Find out the largest size of steelwork that can be powder-coated, and gather some pictures of examples.

Effects of adverse weather on building materials used in domestic and commercial dwellings

Driving rain, snow and wind

Water can be driven into a structure and this can occur in several ways: through the roof tiles and onto the underlay, through open windows or poor seals or through badly maintained brickwork mortar joints. This can lead to damage. The recent spate of floods within the UK have shown just what damage can occur to a property even if the water did not enter the ground floor. Excessive rising damp can occur which causes damage to internal finishes, timber wall plates and floor timbers if the water contains sewage particles. High volumes of floodwaters can cause buildings to move on their foundations from the water pressure or can, in fact, push over walls.

High levels of snowfall can produce heavy loads on structures that could lead to cracking or collapse of the roof structure. High winds each autumn also account for some damage to chimneys and walls as brickwork is blown over.

Knowledge refresher

1 Identify a sustainable material with reasons for your choice.

2 How would you identify if a material is sustainable?

3 What effect does water have on timber?

4 Does anything have to be done with brickwork panels for movement?

5 Why would timber be chemically treated?

6 What method could be used to protect mild steel?

7 What does fire do to steel beams?

chapter 7

Advanced first fixing

OVERVIEW

Carpenters and joiners will carry out many different types of work during their career, and this chapter looks at the first fix aspect of that work. First fixing is work that is usually done before the plastering, including:

- windows
- stairs
- joist and stud coverings
- temporary work and formwork.

All of the above are covered comprehensively in *Carpentry and Joinery NVQ and Technical Certificate Level 2* Chapter 9. In this book we will look more closely at two of these areas of work, as well as introducing two new topics.

The areas we will be looking at more closely are:

- windows
- stairs.

The two new topics this chapter will cover are:

- joist and stud coverings (plasterboard)
- temporary work and formwork.

Windows

Windows are designed with three main factors in mind:

- allowing natural light into the room/dwelling
- protecting the room/dwelling from weather
- keeping the heat in and cold out of a room/dwelling.

With windows it is also important to take both planning and conservation into consideration. Planning is vital when installing new windows as in some cases the type and style of window being fitted must be the same as the ones being replaced – this is especially important when working on listed buildings and within conservation areas.

Window parts

Different types of windows have different parts, but the majority will have:

- **sill** – the bottom horizontal member of the frame
- **head** – the top horizontal member of the frame
- **jambs** – the outside vertical members of the frame
- **mullion** – intermediate vertical member between the head and sill
- **transom** – intermediate horizontal member between the jambs.

If a window has a sash, the sash parts will consist of:

- **top rail** – the top horizontal part of the sash frame
- **bottom rail** – the bottom horizontal part of the sash frame
- **stiles** – the outer vertical parts of the sash frame
- **glazing bars** – intermediate horizontal and vertical members of the sash frame.

There are various different styles of window, the most common types being:

- **fixed glazing** – a frame with no opening sashes with the glass fixed directly into it
- **box frame or sliding sash window** – a window with sliding sashes
- **casement window** – a window with sashes that are hung either on the top or side and can be traditional or storm proof
- **bay window** – window that projects out from the building
- **bow window** – similar to bay windows but more segmental and rounded.

Definition

Listed building – a building that the national heritage body or local council has defined as protected. Any work done to a listed building is stringently monitored and controlled by the planning authority

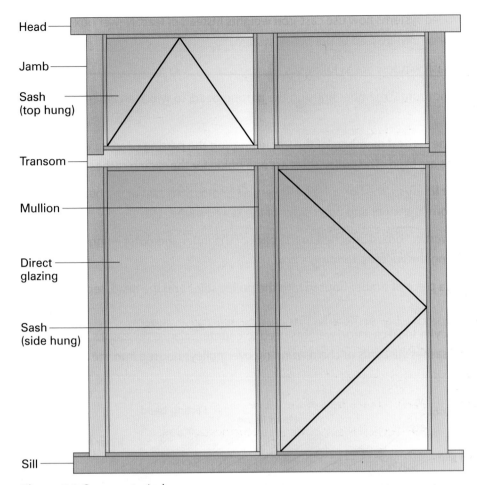

Figure 7.1 Casement window

Casement windows are covered in *Carpentry and Joinery NVQ and Technical Certificate Level 2* and fixed glazing is simply a frame with a piece of glass fitted into it, so here we will concentrate on the box frame, bay and bow windows.

Box frame sliding sash window

Often referred to as a box sash window, this is rarely used nowadays – the casement window is preferred as it is easier to manufacture and maintain. Box sash windows are mainly fitted in listed buildings or in buildings where like-for-like replacements are to be used. Sometimes box sash windows are fitted because the client prefers them.

The box sash window is constructed with a frame and two sashes, with the top sash sliding down and the bottom sash sliding up. Traditionally the sashes work via pulleys, with lead weights attached to act as a counterbalance. The weights must be correctly balanced so that the sashes will slide with the minimum effort and stay in the required position: for the top sash, they must be only slightly heavier than the weight of the glazed sash; for the bottom sash, slightly lighter than the weight of the glazed sash.

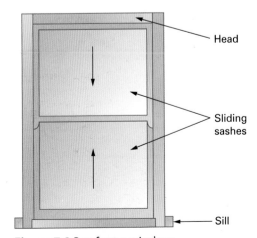

Figure 7.2 Box frame window

Most pulley-controlled systems now use cast-iron weights instead of lead. Newer box sash windows use helical springs instead of the pulley system.

The parts of a box sash window are as follows:

- **head** – the top horizontal member housing the parting bead, to which an inner and outer lining are fixed

- **pulley stiles** – the vertical members of the frame housing the parting bead, to which inner and outer linings are fixed. The pulley stiles act as a guide for the sliding sashes

- **inner and outer linings** – the horizontal and vertical boards attached to the head and pulley stiles to form the inner and outer sides of the boxed frame

- **back lining** – a vertical board attached to the back of the inner and outer linings on the side and enclosing the weights, forming a box

- **parting bead** – a piece of timber housed into the pulley stiles and the head, separating the sliding sashes

- **sash weights** – cylindrical cast iron weights used to counterbalance the sashes. These are attached to the sashes via cords or chains running over pulleys housed into the top of the pulley stiles

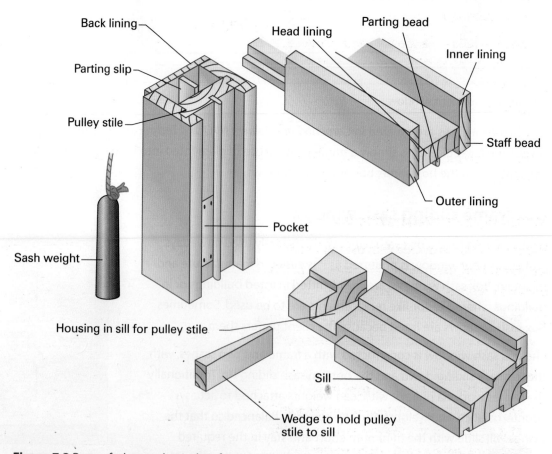

Figure 7.3 Parts of a box sash window frame

- **sill** – the bottom horizontal part of the frame housing
the pulley stiles and vertical inner and outer linings

- **staff bead** – a piece of timber fixed to the inner lining, keeping the lower sash in place

- **parting slip** – a piece of timber fixed into the back of the pulley stiles, used to keep the sash weights apart

- **pockets** – openings in the pulley stile that give access to the sash cord and weights, allowing the window to be fitted and maintained.

The sash in a box frame is made slightly differently to normal sashes, and has the following parts:

- **bottom rail** – the bottom horizontal part of the bottom sash

- **meeting rails** – the top horizontal part of the bottom sash and the bottom horizontal part of the top sash

- **top rail** – the top horizontal part of the top sash

- **stiles** – the outer vertical members of the sashes.

Installing a box sash window

This can be done in one go with the sashes and weights already fixed, but for the purposes of this book we will show a more traditional method.

1. First strip the frame down and have the sashes, parting bead, staff bead and pockets removed.

2. Then fix the frame into the opening, ensuring that it is both plumb and level, and fit the sash cord and sashes (the fitting of the sash cord to the sashes will be covered in Chapter 10 page 278).

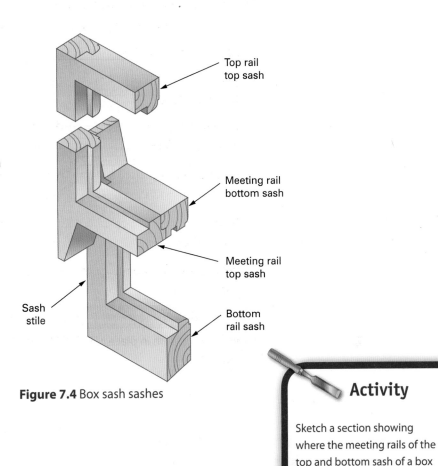

Top rail top sash

Meeting rail bottom sash

Meeting rail top sash

Sash stile

Bottom rail sash

Figure 7.4 Box sash sashes

Activity

Sketch a section showing where the meeting rails of the top and bottom sash of a box sash window meet.

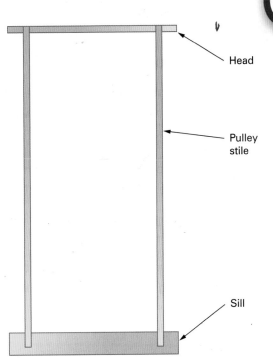

Head

Pulley stile

Sill

Figure 7.5 Stripped down frame

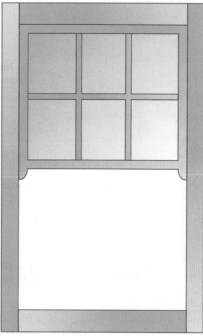

Figure 7.6 Top sash in place

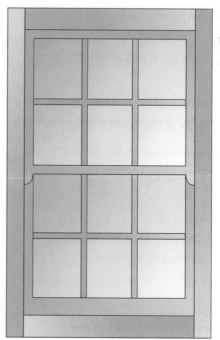

Figure 7.7 Top and bottom sashes in place

3. Next slide the top sash into place and fit the parting bead, to keep the top sash in place.

4. Now fit the bottom sash, then the staff bead, to keep the bottom sash in place.

5. Finally, seal the outside with a suitable sealer, fit the ironmongery and make good to the inside of the window.

The marking out of windows will be covered in Chapter 12 and the maintenance of a box sash window in Chapter 10.

Bay window

Bay windows project out from the face of the main outer walls of the dwelling. There are several different styles, but only three ways of building or incorporating one into a house.

Method 1 Pre-built bay

This is the simplest method, as it is part of the original design of the house and is constructed when the base walls of the house are. The area in the bay contributes to the size of the interior of the room.

Method 2 Built bay

With this method a bay is added onto the original structure of the house. The builder may open up the existing wall and build the bay into it, giving a larger room, or leave the wall and have a large window board area that can be used for storage – the choice depends on both the bay size and the client's wishes.

Method 3 Supported bay

This is only used for small bays with a small projection. Here a support system carries the weight of the bay, but no wall is built. The room is no bigger, but there will be a larger window board area.

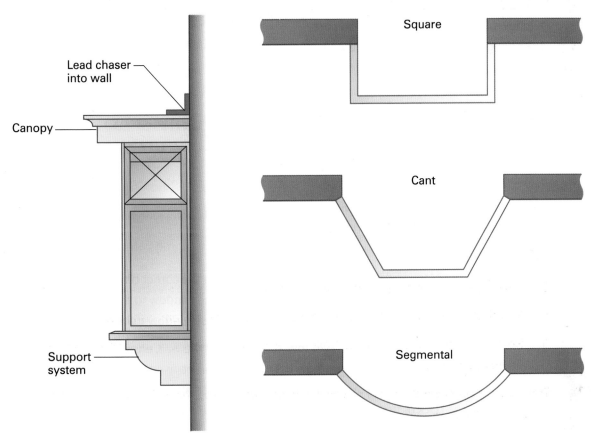

Figure 7.8 Supported bay

Figure 7.9 Common types of bay window

Types of bay window

As you will know from Level 2, several types of bay window can be constructed. The choice of style will depend on the size and shape of the opening or existing bay.

Bay windows are usually made up as a series of windows joined together rather than a single window. Bay windows are joined together by either manufacturing the windows to allow them to be screwed together or by using pre-made posts.

If the window is a square bay, the windows can be joined together like this:

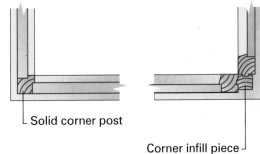

Figure 7.10 Square bay joints

If the window is a splayed or cant bay, they can be joined together as shown in Figure 7.11.

There may need to be a joint along the front of a bay window (usually only on very large windows), in which case the windows are joined together as shown in Figure 7.12.

Installing a bay window

The first thing to do is to remove the old window. The old window will be load bearing, so the roof above it must be suitably supported before it is removed.

Next, fit and level the windowsill. Then fit one of the end windows, level and plumb it, then fix it to the wall.

Solid corner post — Corner infill piece

Figure 7.11 Cant bay joints

Cover strip

Infill piece

Figure 7.12 Lengthening bay joints

Now fit the corner block and the front window, screwing it to the corner block and ensuring it is level, plumb and square to the first window. Do not fix the window to the sill or roof/wall above yet as it may need to be moved slightly to allow the last window to be fitted.

Next fit the final corner block and last window, screwing it to the wall, ensuring it is level, plumb and square to the front window. Now the windows can be fixed to the sill and roof/wall above and, if required, the cover strips at the joints can be fitted. Running a small bead of silicone along the back of the cover strip prior to fixing will help with waterproofing.

Figure 7.13 Fit the windowsill

Figure 7.14 Fit the first window

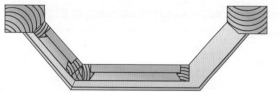

Figure 7.15 Fit the front window

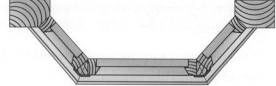

Figure 7.16 Window fitted including cover strips

To finish, glaze the window, fit any ironmongery, seal the exterior with silicone, fit the window board and make good to the inside.

Remember

When removing an old window, the less damage you do, the less you will have to make good later.

Remember

When fitting the first window, it is a good idea to run a small bead of **silicone** between the windows and corner posts to help with waterproofing.

Activity

In an area designated by your trainer, install a bay window.

Bow window

Bow windows are similar to segmental bay windows. They are made as a single window if the radius of the arch is not too great; otherwise they are made as separate windows.

The segmental bow window is fitted in the same way as the bay window, with the sill fitted first and the windows fitted from one side to the other. A bow window made as a single window is fitted in the same way as a normal window, with extra care being taken when handling the large window.

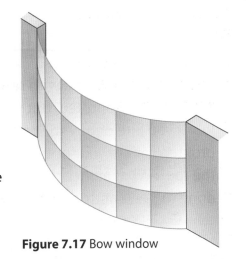

Figure 7.17 Bow window

Activity

In an area designated by your trainer, install a bow window.

Knowledge refresher

1   What are the three main factors to take into consideration when windows are designed?

2   Name the four main parts of a window frame.

3   Name the three main parts of a window sash.

4   Name three main types or styles of window.

5   What type of weights is used in a box sash window?

6   How heavy should the weights balancing the top sash in a box sash window be?

7   Name three different types of bay window.

8   What is the purpose of a staff bead in a box sash window?

What would you do?

You have been tasked with replacing a bay window. The upstairs should be properly supported, but you forgot to price the bracing into the job: if you hire or buy the correct equipment, it will seriously affect any profit. What are your options? What could happen if you do not brace the roof? What alternatives to bracing are there? What could have been done to prevent this problem?

Stairs

There are various different types of stairs, including:

- **straight flight** – where the stairs go straight from one floor to another (used where there is plenty of space)

- **quarter turn/quarter landing** – where the staircase turns 90 degrees while rising. This consists of two straight flights meeting at a platform or landing (usually in the middle). This type of stair is used where there is not much space or where the top landing needs to be in a different direction to the bottom flight of stairs

- **dog leg/half-landing/half-turn stair** – where the staircase turns 180 degrees while rising. This stair can be constructed in two ways, the choice depending on the space available or the number of landings required

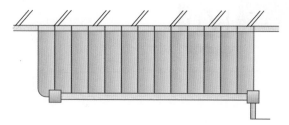

Figure 7.18 Straight flight

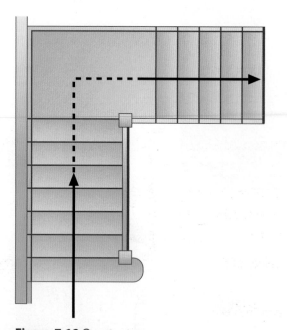

Figure 7.19 Quarter turn

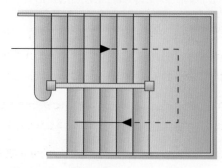

Figure 7.20 Dog leg

- **quarter/half turn with winders** – continually rising with tapered steps fitted at the turn but no platforms. This is used when there is a need for a direction change but no space for a landing

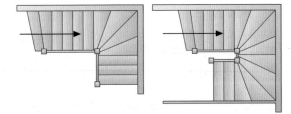

Figure 7.21 Winders

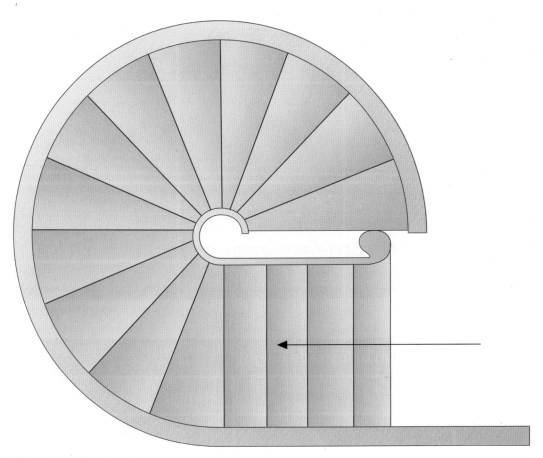

Figure 7.22 Spiral staircase

- **spiral staircase** – completely circular, turning continuously while rising. This is used when there is very little room.

The straight flight is covered fully in *Carpentry and Joinery NVQ and Technical Certificate Level 2 2nd edition* page 250–256, while spiral staircases are usually metallic and fitted by specialists, so here we will look at the set-up of a turn in stairs and the use of winders within a turn.

Before we look at these staircases in more detail it is important to remind ourselves of the regulations that govern staircases, which will be familiar to you from Level 2.

For a list of components and their definitions, see *Carpentry and Joinery NVQ and Technical Certificate Level 2 2nd edition,* pages 250–251.

Activity

Using sketches, show two different ways of creating a 180° turn in a stair.

Description of stair	Max rise (mm)	Min going (mm)	Range (to meet pitch limitation) (mm)
Private stair	220	220	155–220 rise with 245–260 going or 165–200 rise with 220–350 going
Common stair	190	350	155–190 rise with 240–320 going
Stairway in institutional building (except stairs used by staff)	180	280	
Stairway in assembly area (except areas under 100m^2)	180	250	
Any other stairway	190	250	

Table 5.1 Regulations for rise and going

Description of stair	Minimum balustrade height	
	Flight	Landing
Private stair	840	900
Common stairway	900	1000
Other stairway	900	1100

Table 5.2 Regulations for minimum balustrade heights

Description of stair	Minimum width (mm)
Private stair giving access to one room only (except kitchen and living room)	600
Other private stair	800
Common stair	900
Stairway in institutional building (except stairs used only by staff)	1000
Stairway in assembly area (except areas under 100m square)	1000
Other stairway serving an area that can be used by more than 50 people)	1000
Any other stairway	800

Table 5.3 Regulations for width of stairs

Quarter/half-turn staircase
Measuring out for a quarter-turn staircase

Look at this as two separate staircases. The first runs from the floor to the landing, and the second from the landing to the destination (top landing). When measuring out, it is important to get the landing sizes correct for the staircase to fit.

There are two methods of measuring out for and fitting a quarter-turn stair.

Method 1

The landing is built first, then each staircase is measured separately. The measurements are then sent to a joiner's shop and the staircases are made. The made staircases are delivered and fitted, with the handrails and **balustrade** fitted at the same time. This method is often preferred as there is less chance of making a mistake.

Method 2

Here all the measurements are taken first and sent to the joiner's shop, so that the stairs can be made while the landing is built. Once delivered, the stairs can be fitted along with the handrails and balustrades. If the measurements are not accurate, this method can lead to mistakes.

Activity

In an area designated by your trainer, install a flight of stairs with winders.

Definition

Balustrade– unit comprising handrail, newels and the infill between it and the string, which provides a barrier for the open side of the stair

Fitting a quarter-turn staircase

For this example, we will use the first method.

First build the landing to the correct dimensions and in line with Building Regulations. The landing should be built out of minimum 4″ × 2″ softwood timber, and ideally be constructed as shown.

Figure 7.23 Landing

Next fit the staircases (bottom staircase first).

Step 1 Fix the wall string, cutting it off at floor level to suit the skirting height.

Step 2 Use a handsaw to cut the plumb cut on the string at the foot of the stairs.

Step 3 The staircase is now level to the floor.

Step 4 Steps 1 and 2 can be repeated at the top with the underside cut out to sit on to the floor trimmer and the top tread is cut away so that it sits on the trimmer.

Step 5 Fit the second staircase in the same way except that the cut at the bottom of the landing should be notched to sit on the floor.

Finally fit the newel posts, handrails, etc.

Step 6 Mortise the outer string into the newel posts at each end.

Step 7 Fix the newel posts in place. The bottom newel can be held in position using various methods depending on the composition of the floor. For rigidity and strength, the top newel should be notched over the trimmer joist and screwed or bolted to it.

Step 8 Fix the wall string to the wall in approximately four places below the steps, unless access is difficult usually with 75 mm screws and plugs. The balustrades and handrail can be fitted once the stairs are secure.

Step 9 Once the staircase has been fitted, it should be protected to prevent damage. Strips of hardboard should be pinned to the top of each tread with a lath to ensure the nosing is protected. Use the same method to protect the newel posts.

A half-turn staircase should be built the same way, preferably with the landing being built first.

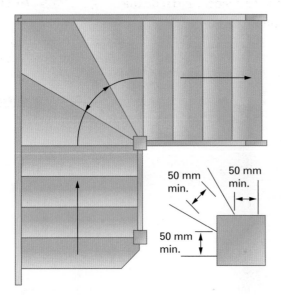

Figure 7.24 Plan drawing of stairs

50 mm min.

50 mm min.

50 mm min.

The use of winders in a turn

'Winders' are the tapered treads on a staircase that turns as it rises without using a landing. The first thing to note is that there are extra regulations regarding winders.

- The rise of the tapered steps must be the same as the rise of the other steps.

- The tapered step must not be less than 50 mm at its narrowest point.

- The going of the tapered steps (measured at the centre of the steps) must be the same as the going of the other steps.

Setting out a staircase containing winders

The number of winders in a tapered stair with a quarter or half turn depends on the size of the staircase. A staircase that is wide with a large going will have more winders than a small staircase. Our example will use a three-winder staircase.

First do a scale plan drawing of the stairs, to help you determine the size and going of the winders.

Next comes the construction. This is fundamentally the same as for a straight flight (covered in *Carpentry and Joinery NVQ and Technical Certificate Level 2*), but with a few differences:

- the **strings** are shaped differently, as the wall strings need to be larger to allow for the rise during the change of direction. Figure 7.25 gives a typical example of how the wall strings will look

- the **newel post** has to be housed out accordingly as well (see figure 7.26).

Now fit the stairs. The made staircase should be delivered to site in two separate parts, with the newel post and winders taken out (to allow for ease of transportation and prevent damage). The simplest way to fit it is to treat it as two separate quarter-turn stairs.

Figure 7.25 Wall strings for winders

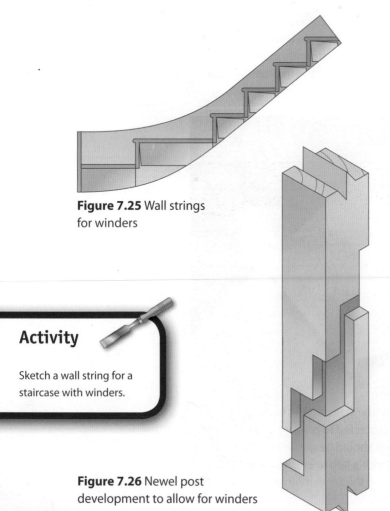

Figure 7.26 Newel post development to allow for winders

Activity

Sketch a wall string for a staircase with winders.

Fix the top section in place, ensuring that the cuts at the top are correct and the treads are level. Fit the newel post to support the open end of the stair. Position the winders and slide the bottom section into place. Check and adjust for level, then fix the bottom section, handrails and balustrades.

Activity

In an area designated by your trainer, install a straight flight of stairs.

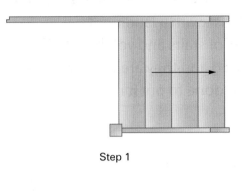

Step 1

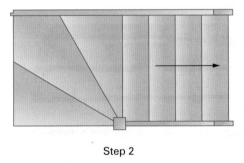

Step 2

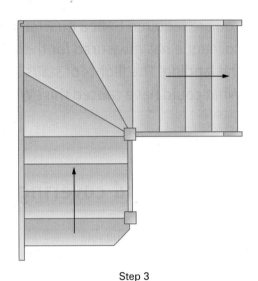

Step 3

Figure 7.27 Three stages of winders being fitted

Knowledge refresher

1   Name two ways in which a turn in stairs can be achieved.

2   What are the minimum rise and going for a private stair?

3   What are the maximum rise and going for a common stair?

4   What is the minimum balustrade height for a private stair?

5   State the minimum width of a common stair.

6   When using winders, what is the minimum width of the winder at its smallest point?

7   Describe how stairs can be protected after installation.

What would you do? !

You are required to fit a quarter-turn staircase with a landing. A colleague has measured for the two flights of stairs and you have built the landing. The flights of stairs are delivered and they are both too big. What could have caused this? What can be done to put it right?

Joist and stud coverings

Joists are decked with a suitable flooring material, and the underside of the joists is usually covered with plasterboard. Stud partitions are also usually covered with plasterboard, with the occasional exception. For example, areas such as shower rooms are likely to be tiled, so a sturdier board such as WBP plywood is used.

Joist and stud coverings have come a long way in the past few years, and the traditional method of lath and plaster is now only really used in listed buildings.

Lath and plaster

There are two main ways of plasterboarding a room:

Method 1

The plasterboard is fixed to the stud or joist with the back face of the plasterboard showing. The plasterer covers the whole wall or ceiling with a thin skim of plaster, leaving a smooth finish.

Plastered wall

Taped wall

Method 2

The plasterboard is fixed with the front face showing, and the plasterer uses a special tape to cover any joints, then a ready mixed filler is applied over the tape and is used to fill in the nail and screw holes. Once the taped area is dry, the plasterer then gives the area a light sand to even it out.

Plasterboard comes with a choice of two different edges, and the right edge must be used:

- square edge – for use with Method 1, the whole wall/ceiling is plastered
- tapered edge – for use with Method 2, so that the plasterer can fix the jointing tape.

As well having the correct edging, the right type of plasterboard must be used.

Various types of plasterboard are available so a colour coding system is used (NB this may vary from manufacturer to manufacturer):

- fire resistant plasterboard (usually red or pink) – used to give fire resistance between rooms such as party walls
- moisture resistant plasterboard (usually green) – used in areas where it will be subjected to moisture such as bathrooms and kitchens

- vapour resistant plasterboard – with a thin layer of metal foil attached to the back face to act as a vapour barrier, usually used on outside walls or the underside of a flat roof

- sound resistant plasterboard – used to reduce the transfer of noise between rooms and can also help acoustics, used in places such as cinemas

- thermal check board – comes with a foam or polystyrene backing, so used to prevent heat loss and give good thermal insulation

- tough plasterboard – stronger than average plasterboard, it will stand up to more impact, used in public places such as schools and hospitals.

Sometimes plasterboard is doubled up – the walls are 'double sheeted' – to give fire resistance and sound insulation without using special plasterboard.

As well as a variety of types plasterboard comes in a variety of sizes, though the most standard size is 2400 × 1200 × 12.5 mm.

Cutting plasterboard

Plasterboard can be cut in two main ways: with a plasterboard saw or with a craft knife. The plasterboard saw can be used when accuracy is needed but most jobs will be touched up with a bit of plaster, so accuracy is not vital. Using a craft knife is certainly quicker.

This is how to cut plasterboard with a craft knife:

Step 1 Mark the line to be cut with a pencil.

Step 2 Score evenly along the line with the craft knife ensuring hands, legs, etc. are out of the way of the blade.

Step 3 Turn the board round and slap the area behind the cut, to split the board in two.

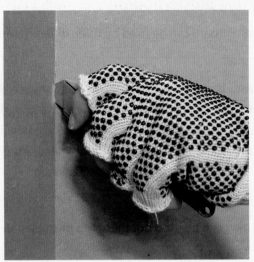

Step 4 Run the knife along the back of the cut to separate the two pieces of board.

Step 5 Trim the cut with a slight back bevel to give a neater finish.

Fixing plasterboard

Plasterboard can be fixed by three different methods, which are:

- **nail**

Plasterboard can be fixed using clout or special plasterboard nails. When nailing, make sure the nail is driven below the surface with the hammer, leaving a dimple or hollow that can be filled by plaster later. The nails used must be galvanised: if not, when the plaster is put on to fill the hollows, the nails will rust and will 'bleed' – the rust from the nails will stain the plaster and can even show through several coats of paint.

- **screw**

Screws can be used but these too must be galvanised to prevent bleeding.

- **dot and dab**

This is used where the plasterboard is fitted to a pre-plastered or flat surface such as block work. With dot and dab, there is no need for a stud or frame to fix the plasterboard, thus increasing the room size (it is often used in stairwells where space is limited). Dot and dab involves mixing up plaster and dabbing it onto the back of the board, then pushing the board directly onto the wall.

Activity

In an area designated by your trainer, cut and fit plasterboard to a timber stud partition.

Remember

Dot and dab is a skilled technique, as with this method getting the boards plumb and flat can be difficult.

Safety tip

When fixing plasterboard using nails or screws, be aware of where the services are. Otherwise, you could hit an electricity cable or a water pipe.

Activity

In an area designated by your trainer, cut and fit plasterboard using dot and dab.

Knowledge refresher

1   When covering studs, when would plywood be used rather than plasterboard?

2   What is the difference in use between square-edge and tapered-edge plasterboard?

3   List four different types of plasterboard and their uses.

4   What is the standard size of plasterboard?

5   What is dot and dab?

6   Why must galvanised nails or screws be used when fixing plasterboard?

What would you do?

You have been tasked with plaster boarding a stud partition wall. You are on the last sheet when you accidentally nail a water pipe. What could the consequences be? What could the consequences be if you had hit a different service (gas, electric)? What could have been done to avoid this?

Temporary work and formwork

As the name suggests, temporary work is work that is only there for a short time and will be removed prior to the finish. There are various types of temporary work and formwork such as:

- **support work** – can take several shapes such as arch centres, which are used to support brick arches

- **shuttering** – where a frame is built, into which concrete is poured. Shuttering is used on jobs like concrete columns and pillars

- **shoring** – where an existing building or component is supported to prevent falling during modification work. For example, where a ceiling or roof is supported, when a lintel or load-bearing partition is being moved or modified, etc.

- **mould boxes** – a timber box is made to a specific shape, into which concrete is poured. Once the concrete has set the box is removed, leaving a pre-cast member such as a windowsill or even a staircase.

We will now look at an example of each of these.

Support work

This is where something is made to support the work while it progresses, a perfect example being an arch centre. An arch centre is used to support the brickwork while the arch is being built, and is usually left *in situ* until the keystone is fitted and the cement has set.

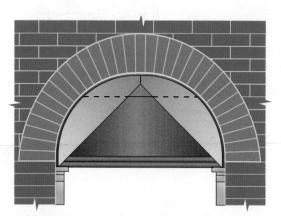

Figure 7.28 Brick arch with arch centre

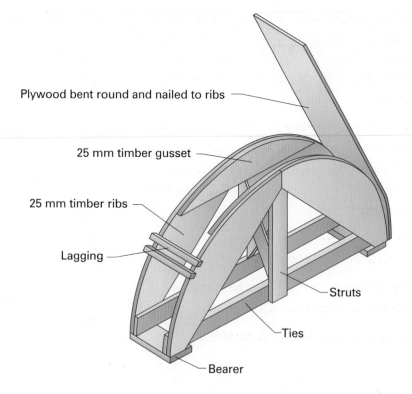

Plywood bent round and nailed to ribs

25 mm timber gusset

25 mm timber ribs

Lagging

Struts

Ties

Bearer

Figure 7.29 Arch parts

Arch centres are usually made from timber. If the span of the opening is larger than 1.5m, a stronger metal centre is used. The arch centre has several parts:

- **ribs** – usually made from 19 mm plywood, MDF or 22 mm solid timber, these form the outer part of the centre, giving the outline for the rest of the framework

- **ties** – horizontal members attached at the bottom of the ribs to prevent the load pushing the ribs out of shape

- **struts** – vertical members used to transfer the weight and spread the load

- **bearers** – flat members fixed to the bottom of the ribs to tie the ribs together

- **bracing** – timber fixed at an angle on the inside of the arch to give support and keep the arch square

- **laggings** – either thin sheets of hardboard or plywood nailed across the top of the ribs, or small timber laths nailed at right angles to the ribs.

When setting out for an arch centre you need to know the span, arch radius and so on. You can get this information from the bricklayer or from drawings.

First draw out the arch full size. Use the drawing to make templates out of hardboard, which can then be traced onto the ribs. The way the ribs are made up depends on the size and radius of the arch: some arches use two ribs per side while others use four.

Now nail the ribs together using gusset plates, fixing them where the gusset overlaps the ridge, then fix the struts and ties. Usually both ribs are made as a pair and checked together against the drawing made earlier.

The bracing is fixed next, bringing both ribs together, then the bearers can be fitted.

Finally fit the lagging, which is nailed down through the tops of the ribs.

Now the arch centre is ready to be used.

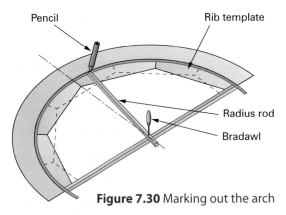

Figure 7.30 Marking out the arch

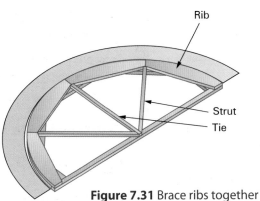

Figure 7.31 Brace ribs together

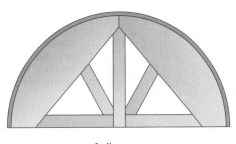

2-rib set up

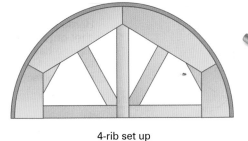

4-rib set up

Figure 7.32 Two-rib and four-rib set-ups

Remember

When marking out the ribs, you must allow for the thickness of the lagging otherwise the arch will be too big.

Activity

In an area designated by your trainer, build an arch centre.

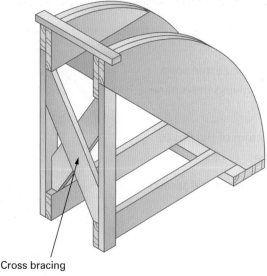

Figure 7.33 Fit bracing and bearer

Cross bracing

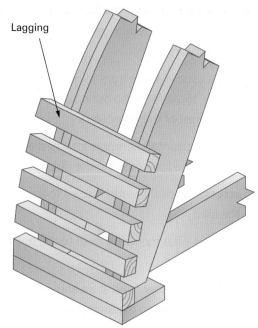

Lagging

Figure 7.34 Fit lagging

Definition

Upstand or **kicker** – a concrete cast base, which is used to help locate a pillar of column in the correct place

Ballistic tool – any of a range of tools where the mechanism involves a 'throwing' action

Shuttering

Shuttering is used mostly on columns and pillars. The method for shuttering a column is as follows:

First form an **upstand** (or **kicker**) for the column. This helps locate the position of the column and reduces the amount of concrete lost at the base of the column when it is poured.

The kickers are formed using a frame fixed to the floor, which is filled with concrete, and a **ballistic tool**. When casting the kickers it is vital that they are in the correct position.

Next fit the shuttering round where the column will go. The shuttering is made up of a series of timber frames clad on one side with plywood, and must be fixed together with screws.

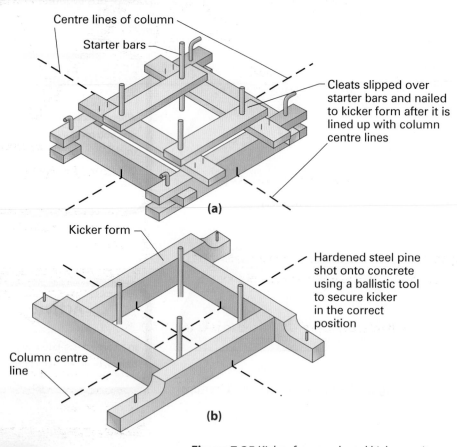

Centre lines of column

Starter bars

Cleats slipped over starter bars and nailed to kicker form after it is lined up with column centre lines

(a)

Kicker form

Hardened steel pine shot onto concrete using a ballistic tool to secure kicker in the correct position

Column centre line

(b)

Figure 7.35 Kicker formwork and kicker on its own

Now fit a series of clamps to the shutter to prevent the concrete bursting through the sides.

The whole structure must now be levelled, plumbed and adequately braced with diagonal bracing to prevent any movement.

Pour the concrete. Once it has set, remove the bracing, clamps and shuttering.

Remember

The plywood used must be clean and have no holes or gaps through which concrete can escape.

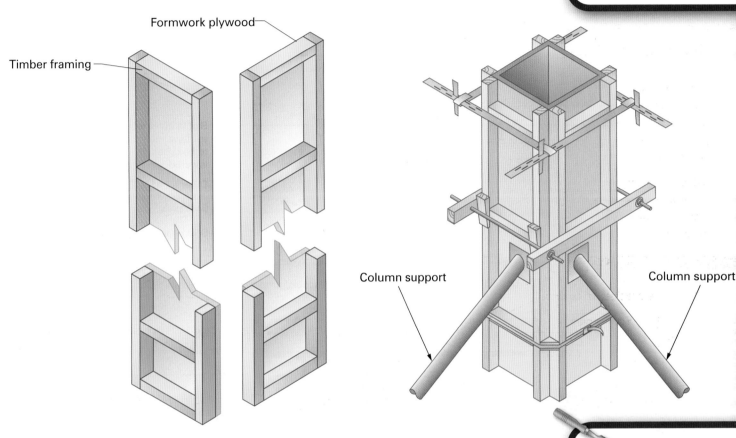

Formwork plywood

Timber framing

Column support

Column support

Figure 7.36 Shuttering

Figure 7.37 Braced column

Shoring

Shoring is used where an existing building or component is supported to prevent falling or collapse during modification or removal work. Our example here is the removal of a load-bearing stud partition.

Proper shoring is vital, and is most commonly done with **telescopic** props adjusted to take the weight of the floor above.

Activity

In an area designated by your trainer, erect shuttering for a column.

Definition

Telescopic– sliding or arranged like the joints of a telescope, so it can be packed away small

Removal of a load-bearing partition

First determine which way the joists on the upper floor run, to make sure your props will support the weight.

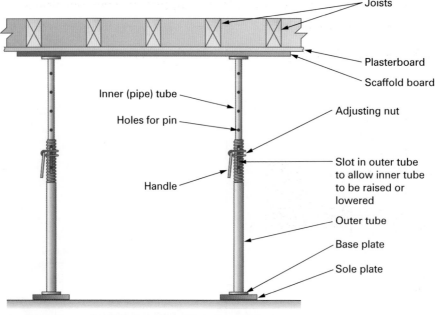

Correct way of positioning adjustable (telescopic) prop

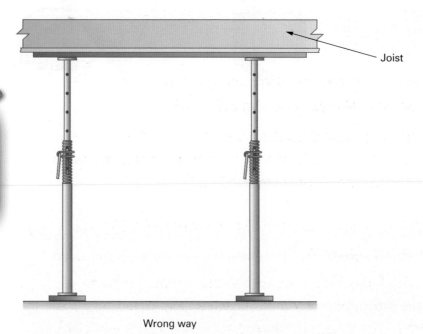

Wrong way

Figure 7.38 Right and wrong way to set up props

Ideally, the props should be placed as close to the partition as possible without being in the way of operations, with the upper floor propped on both sides of the partition. The props should be placed onto a board, acting as a sole plate, and a board should also be positioned where they meet the ceiling, to spread the load.

Now the work can commence. The props should only be removed once the work is complete and a suitable support system such as a steel beam is in place.

Mould boxes

Mould boxes are similar to shuttering, but are used to form items that are more complicated. Mould boxes can be dismantled, re-assembled and used again.

A simple mould box for a windowsill would look something like Figure 7.39.

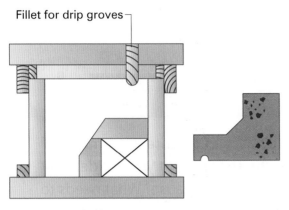

Fillet for drip groves

Figure 7.39 Windowsill mould box

Remember

When making mould boxes, the mould needs to be the *reverse* of the finished component, otherwise the component will not be right.

FAQ

I didn't know there were so many different types of plasterboard available. Which type should I use?

The type of plasterboard used should be stated by the client or in the architect's drawings. There are different types because there are so many different purposes that rooms are used for.

I have been told that I have nailed the plasterboard on the wrong way. Which way is the right way?

Plasterboard has two faces. The front face is usually white with no markings, while the back has the manufacturer's logo on it. The right way will depend on what is happening to the plasterboard. If it is only having the joints taped, then the white face must be shown; if the whole wall is being skimmed, the plasterers usually prefer to have the back face showing.

Knowledge refresher

1   What is the main purpose of support work?

2   What is the main purpose of shuttering?

3   What type of plywood should be used for shuttering?

What would you do?

You have been tasked with shoring up a ceiling before the removal of a wall to form a doorway. You set up the shoring, but you accidentally overtighten the props. What damage could this cause? What needs to be done to rectify the damage? How do you determine how tight the props should be?

OVERVIEW

Second fixing covers all the carpentry work done after the plastering. This work has to be more precise than first fixing work as it is usually visible. Second fixing work involves:

- mouldings (skirting, architrave, etc.)
- door hanging
- ironmongery (screws, nails locks, etc.)
- encasing services (pipe boxes, bath panels, etc.)
- wall and floor units (kitchens, bathrooms, bedrooms, etc.).

All of the above are covered in *Carpentry and Joinery NVQ and Technical Certificate Level 2*, so the aim here is to expand your knowledge of some areas and to look at more complex aspects of the work.

The topics we will be looking at are:

- external doors (including ironmongery)
- hanging an external door
- double doors (including ironmongery)
- sliding doors
- dado, picture rail and cornice
- wall panelling.

External doors

This section aims to provide the knowledge and understanding needed to select, hang and fix the required ironmongery to an external door.

Types of external door

External doors are always solid or framed – a flush or hollow door will not provide the required strength or security.

In some cases external doors are made from UPVC. UPVC doors require little maintenance and the locking system normally locks the door at three or four different locations, making it more secure. High quality external doors are usually made from hard-wearing hardwoods such as mahogany or oak, but can be made from softwoods such as pine.

External doors come in the same dimensions as internal doors except external doors are thicker (44 mm rather than 40 mm). On older properties external doors often have to be specially made as they tend not to be of standard size.

There are four main types of external door:

- framed, ledged and braced
- panelled doors
- half/full glazed doors
- stable doors.

Framed, ledged and braced doors

Framed, ledged and braced (FLB) doors consist of an outer frame clad on one side with tongued and grooved boarding, with a bracing on the back to support the door's weight.

FLB doors are usually used for gates and garages, and sometimes for back doors. When hanging an FLB door it is vital that the bracing is fitted in the correct way; if not, the door will start to sag and will not operate properly.

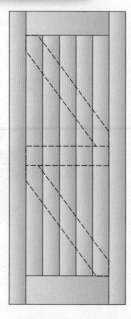
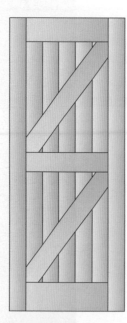

Front elevation   Back elevation

Figure 8.1 Front and back of a framed, ledged and braced door

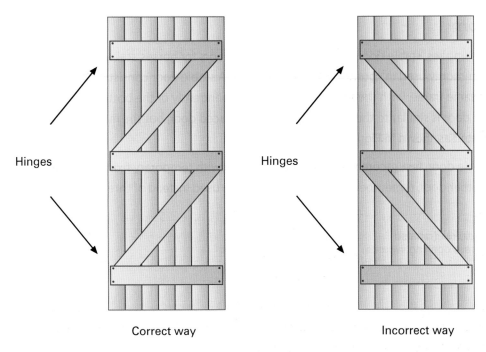

Figure 8.2 Bracing fitted correctly and incorrectly

Panelled doors

Panelled doors consist of a frame made up from stiles, rails, muntins and panels. Some panel doors are solid, but most front and back doors have a glazed section at the top to allow natural light into the room.

Half/full glazed doors

Half-glazed doors are panelled doors with the top half of the door glazed. These doors usually have diminished rails to give a larger glass area.

Full glazed doors come either fully glazed or with glazed top and bottom panels, separated by a middle rail.

Full glazed doors are used mainly for French doors or back doors where there is no need for a letterbox.

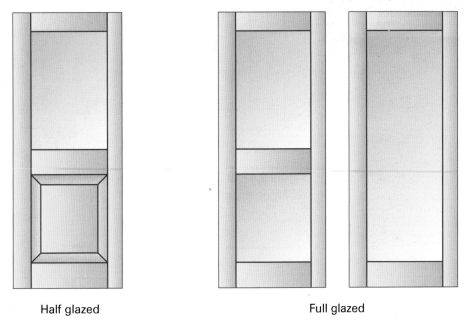

Figure 8.3 Half-glazed door and two types of full glazed door

Stable doors

As their name implies, stable doors are modelled on the doors for horses' stables and they are now most commonly used in country or farm properties. A stable door consists of two doors hung on the same frame, with the top part opening independently of the bottom. The make-up of a stable door is similar to the framed, ledged and braced door, but the middle rail will be split and rebated as shown in Figure 8.4.

To hang a stable door, first secure the two leaves together with temporary fixings, then hang just like any other door, remembering that four hinges are used instead of three.

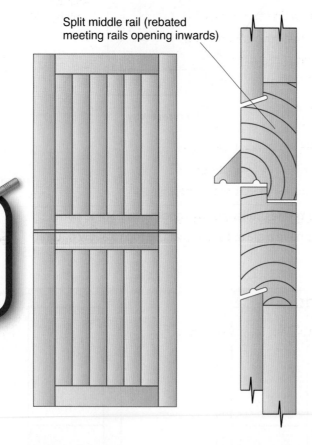

Split middle rail (rebated meeting rails opening inwards)

Figure 8.4 Stable door with section showing the rebate on the middle rails

Activity

In an area designated by your trainer, hang an external stable door.

Hanging an external door

External doors usually open inwards, into the building. Where a building opens directly onto a street, this prevents the door knocking into unsuspecting passers-by, but, even where there is a front garden, having a door that opens outwards is not good practice as callers will need to move out of the way to allow the door to be opened. Externally opening front doors are usually only used where there is limited space, or where there is another door nearby which affects the front door's usage.

Hanging an exterior door is largely the same as hanging an interior door (see *Carpentry and Joinery NVQ and Technical Certificate Level 2 2nd edition* page 292–295) except that the weight of the door requires three hinges sited into a frame rather than a lining. Because of this, there is usually a threshold or sill at the bottom of the doorframe.

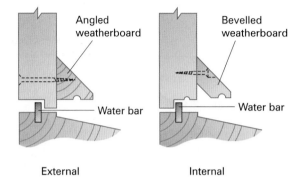

If a water bar is fitted into the threshold to prevent water entering the dwelling, you will need to rebate the bottom of the door to allow it to open over the water bar. The way the door is then hung will depend on which side of the door is rebated. A weatherboard must also be fitted to the bottom of an external door, to stop driving rain entering the premises.

Figure 8.5 Internally and externally opening doors

As an exterior door is exposed to the elements it is important that the opening is draught proofed: a draught proofing strip can be fitted to the frame, or draught proofing can be fitted to the side of the door.

Ironmongery for an external door

An external door requires more ironmongery than an internal door, and may need:

- hinges
- letter plate
- mortise lock/latch
- mortise dead lock
- cylinder night latch
- spy hole
- security chains.

Hinges

External door hinges are usually butt hinges, though framed, ledged and braced doors often use T hinges. Three 4" butt hinges are usually sufficient, though it is advisable to use security hinges (hinges with a small steel rod fixed to one leaf, with a hole on the other leaf) to prevent the door being forced at the hinge side.

Activity

In an area designated by your trainer, hang an external door.

Letter plate

The position of a letter plate depends on the type of door. Letter plates are usually fitted into the middle rail, but could be fitted in the bottom rail (for full glazed doors) or even the stile (with the letter plate fitted vertically).

Activity

In an area designated by your trainer, fit a letter plate.

Did you know?

A lot of draughts can occur in the letter plate area, so it is good practice to have a draught-proofing strip fitted.

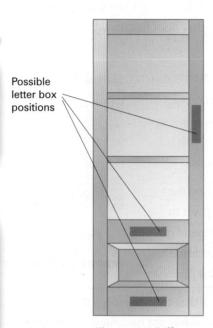

Possible letter box positions

Figure 8.6 Different options for fitting a letter plate

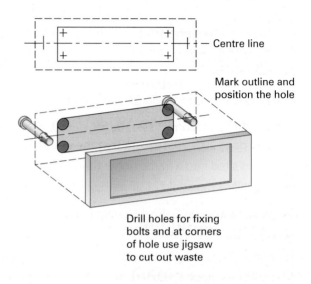

Centre line

Mark outline and position the hole

Drill holes for fixing bolts and at corners of hole use jigsaw to cut out waste

A letter plate can be fitted either by drilling a series of holes and cutting out the shape with a jigsaw, or by using a router with a guide.

Once the opening has been formed, the letter plate can be screwed into place.

Mortise lock/latch

Mortise locks are locks housed into the stile and are fitted as follows:

Step 1 Mark out the position for the lock (usually 900 mm from the floor to the centre of the spindle) and mark the width and thickness of the lock on the stile.

Step 2 Using the correct sized auger/flat bit, drill a series of holes to the correct depth.

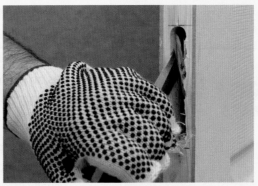

Step 3 Using a sharp chisel remove the excess timber, leaving a neat opening.

Step 4 Mark around the faceplate and with a sharp chisel, remove the timber so that the faceplate sits flush with the stile.

Step 5 Mark where the spindle and keyhole are, then drill out to allow them both to be fitted and operate properly.

Step 6 Fix the lock and handles in place.

Step 7 Mark the position of the striking plate on the doorframe, then house it into the frame.

Step 8 Check that the lock operates freely.

Activity

In an area designated by your trainer, fit a mortise lock.

Basic security ironmongery

A mortise lock on its own does not usually provide sufficient security for an exterior door, so most doors will also have one of the following:

- **mortise deadlock** – fitted like a mortise lock except that it has no latch or handles, so an escutcheon is used to cover the keyhole opening. It is usually fitted three quarters of the way up the door.

- **cylinder night latch** – preferred to a mortise deadlock, as it does not weaken the door or frame as much. It is also usually fitted three quarters of the way up the door. The manufacturer will provide fitting instructions with the lock.

Spy hole

A spy hole is usually fitted to a door that is solid or has no glass, and is used as a security measure so that the occupier can see who is at the door without having to open the door.

To fit a spy hole, simply drill a hole of the correct size, unscrew the two pieces, place the outer part in the hole, then re-screw the inner part to the outer part.

Security chain

A security chain allows the door to be opened a little without allowing the person outside into the dwelling. The chain slides into a receiver, then as the door opens the chain tightens, stopping the door from opening too far.

Double doors

Here two doors are fitted within one single larger frame/lining, with 'meeting stiles': two stiles that meet in the middle, rebated so that one fits over the other.

Double doors are usually used where a number of people will be walking in the same direction, while double swing doors are used where people will be walking in different directions. Double doors allow traffic to flow without a 'bottleneck' effect, and let large items such as trolleys pass from one room to another. Double doors are used in places such as large offices, or public buildings such as schools and hospitals.

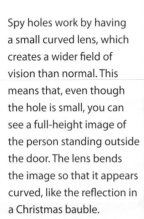

Did you know?

Spy holes work by having a small curved lens, which creates a wider field of vision than normal. This means that, even though the hole is small, you can see a full-height image of the person standing outside the door. The lens bends the image so that it appears curved, like the reflection in a Christmas bauble.

The method for hanging hanging double doors is the same as for any other door, though extra care must be taken to ensure that the stiles meet evenly, and that there is a suitable gap around the doors.

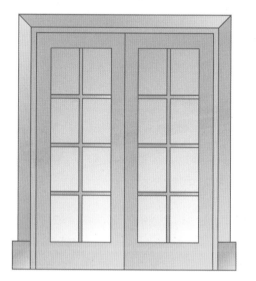

Meeting stiles

Activity

In an area designated by your trainer, hang a set of double doors.

Figure 8.7 Meeting stiles

Extra ironmongery is required on double doors as follows:

- **parliament hinges** – these project from the face of the door and allow the door to open 180 degrees

- **rebated mortise lock** – similar to a standard mortise lock but the lock and striking plate are rebated to allow for the rebates in the meeting stiles

Parliament hinge

Rebated mortise lock

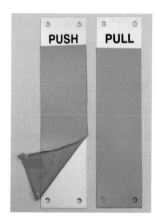

Push plates

- **push/kick plates** – fixed to the meeting stiles and the bottom rails to stop the doors getting damaged

- **door pull handles** – fixed to the meeting stiles on the opposite side from the push plates, these allow the door to be opened

- **barrel bolts** – usually fixed to one of the doors at the top or the bottom to secure the door when not in use.

Pull handles

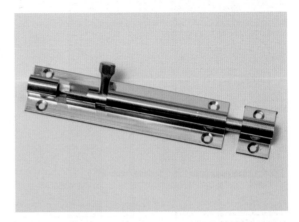

Barrel bolt

Definition

Helical – in a spiral shape

Helical hinge – a hinge with three leaves, which allows a door to be opened through 180 degrees

Most double doors also have door closers to ensure that they will close on their own, to prevent the spread of fire or draughts throughout the building.

There are four main types of door closer: overhead, concealed, floor springs and **helical** spring hinges. Floor springs and **helical hinges** will be looked at on page 214 when we deal with double swing doors, so for the moment we will look at concealed and overhead door closers.

Concealed door closers

Concealed door closers work through a spring and chain mechanism housed into a tube.

To fit a concealed door closer, you must house the tube into the edge of the hinge side of the door, with the tension-retaining plate fitted into the frame.

Overhead door closers

As the name implies, overhead door closers are fitted to either the top of the door or the frame above. They work through either a spring or a hydraulic system fitted inside a casing, with an arm to pull the door closed.

There are different strengths of overhead door closer to choose from depending on the size and weight of the door. Table 6.1 below shows the strengths available.

Power no.	Recommended door width	Door weight
1	750 mm	20kg
2	850 mm	40kg
3	950 mm	60kg
4	1100 mm	80kg
5	1250 mm	100kg
6	1400 mm	120kg
7	1600 mm	160kg

Table 6.1 Strengths of overhead door closures

There are a number of different ways to fit an overhead door closer depending on where the door is situated. The two main ways are:

- fitting the main body of the closer to the door, with the arm attached to the frame

- inverting the closer so that the main body is fitted to the frame and the arm is fitted to the door.

Remember

When choosing an overhead door closer, you must take into account any air pressure from the wind. If the pressure is strong, you may require a more powerful closer.

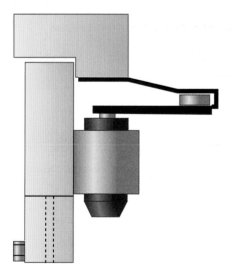

Figure 8.8 Closer fitted to door

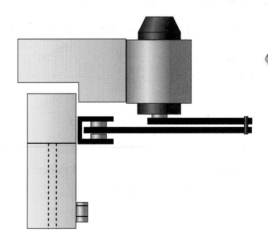

Figure 8.9 Closer fitted to frame

Activity

In an area designated by your trainer, fit an overhead door closer.

Door closers come with instructions and a template, to make fitting easier.

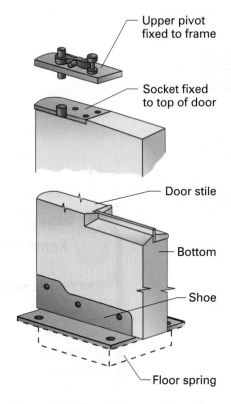

Figure 8.10 Floor spring and top of door fixing

Double swing doors need to open both ways to accommodate traffic going in both directions. The main difference work-wise between standard double doors and double swing doors is in the ironmongery. Double swing doors need special hinges, and must have some form of door closer fitted – and it is usually best to combine these, in one of two ways.

- **Helical spring hinges**

Helical spring hinges are three-leaf hinges with springs integrated into the barrels. The way the leaves are positioned allows the doors to open both ways. The hinges are fitted just like normal hinges and the tension on the springs are adjusted via a bar inserted into the hinge collar.

- **Floor springs**

Doors using floor springs are hung via the floor spring at the bottom and a pivot plate at the top. The floor spring is housed into the floor, and the bottom of the door is recessed to accept the shoe attached to the floor spring.

The pivot plate at the top is attached to the frame, and a socket is fixed to the top of the door.

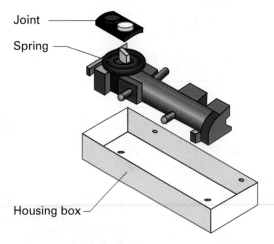

Figure 8.11 Helical floor springs

Sliding doors

Sliding doors are mainly used where there is no space for a door to open outwards. They are not hung in the traditional way with hinges, but use a track on which the door slides open along the face of the wall.

There are many different systems available that allow a door to be fitted in this way, but in most the basics are the same. The door is suspended from an overhead track, and slides on a series of rollers, while the bottom of the door is grooved so that it runs over a track or plastic guide. The overhead track section is usually encased in a pelmet, allowing the door to be detached easily for maintenance.

Remember

Installation differs from manufacturer to manufacturer, so it is best to follow the manufacturer's instructions.

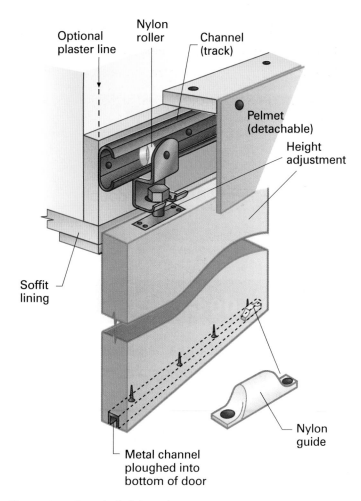

Figure 8.12 Detail of sliding door gear

Activity

In an area designated by your trainer, hang a sliding door.

Knowledge refresher

1 Name the four main types/styles of exterior door.

2 Give an example of where a framed, ledged and braced door would be used.

3 How many hinges are used when hanging a stable door?

4 On an external door, what is the purpose of a water bar?

5 Name five pieces of ironmongery that may be fitted on an external door.

6 State the size, type and number of hinges usually fitted to an external door.

7 What is the purpose of a spy hole?

8 What is the purpose of push plates?

9 Name the four main types of door closer.

10 What is the difference between standard double doors and double swing doors?

What would you do?

1 You have been tasked with fitting a pair of double doors. You fit them but one door is binding on the hinge side, and the doors overlap in the centre at the top, so they do not close properly. What has caused this? What needs to be done to remedy this?

2 You have been tasked with fitting an overhead door closer in a bedroom with windows. You fit the closer and the door closes smoothly until the end, but remains open about an inch. What could be causing this? What can be done to rectify it?

Dado, picture rail and cornice

Carpentry and Joinery NVQ and Technical Certificate Level 2 covered architrave and skirting, so in this book we will expand our knowledge of mouldings to include dado rail, picture rail and cornice.

Mouldings in a room are positioned like this:

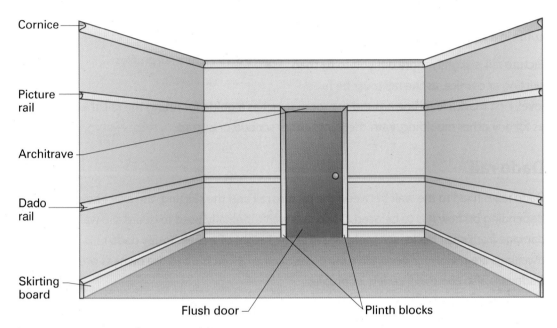

Figure 8.13 Positions of room mouldings

We will start at the top and work down, beginning with the cornice.

Cornice

Cornice is fitted where the top of the wall meets the ceiling. It is traditionally associated with plasterers, although some cornice today is made of timber and is fitted by the carpenter. Cornice can be fitted as a decoration piece or used to hide gaps or blemishes. As with most mouldings, various different designs are available.

Figure 8.14 Cornice profiles

Cornice is simple to install. You nail or screw it into the wall along the ceiling, taking care to ensure that it is running flat to both the wall and the ceiling. External joints are mitred, while internal joints can be either scribed or mitred.

Picture rail

Picture rail is usually fixed at the same height as the top architrave on a door, or just above or below this. Picture rail was used to hang paintings from, so the height used to be determined by the size of pictures being hung; nowadays picture rail is mainly for decoration. Picture rails also come in a variety of profiles.

Figure 8.15 Picture rail profiles

Picture rail is slightly more difficult to fix than skirting or cornice, as the rail must be fixed level. First mark a level line around the room to act as a guide for fitting the rail, then proceed as for any other moulding, with the joints either scribed or mitred.

Dado rail

Dado rail is fixed to the wall between the picture rail and the skirting, the exact height depending on how it is to be used. Dado rails were originally used to guard the walls from damage from chair backs, so would be set at the height of a chair. Today dado rails in domestic dwellings are mainly for decorative purposes, so the height is up to the owner's preference. Dado rails also come in a variety of profiles.

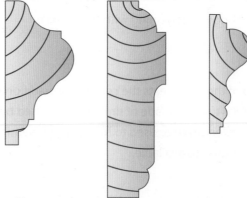

Figure 8.16 Dado rail profiles

You fix a dado rail in the same way as a picture rail, taking care to ensure that the rail is level and the joints used are scribed and mitred.

Wall panelling

Wall panelling provides a decorative finish to a room and can be found in places such as courts of law or executive offices. Wall panelling is usually set at one of three heights:

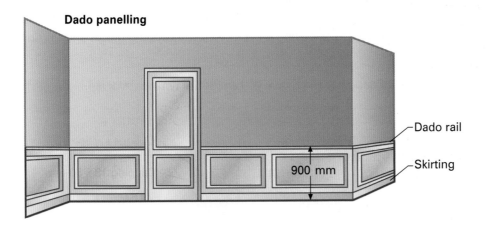

Dado panelling

900 mm

Dado rail

Skirting

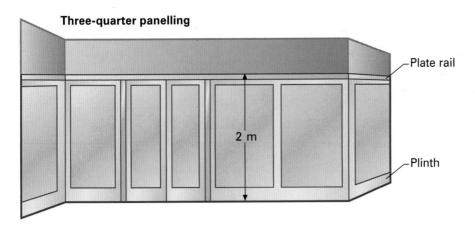

Three-quarter panelling

Plate rail

2 m

Plinth

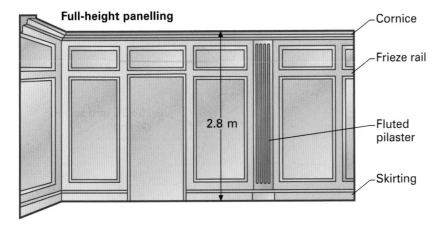

Full-height panelling

Cornice

Frieze rail

2.8 m

Fluted pilaster

Skirting

Figure 8.17 Three typical heights of panelling

- **dado panelling** – where the panelling runs to the height of the dado rail
- **three-quarter panelling** – where the panelling runs to the top of the door
- **full-height panelling** – where the panelling runs from floor to ceiling.

The panelling can be made up in a variety of ways, depending on the type and style of the house. The first thing to do is to make a frame or fix battens to the wall, onto which you can then fix your panelling. Once the panelling is in place, fix the capping pieces and skirting to finish the panelling off.

These three examples of dado panelling give you an idea of what is involved.

Activity

Sketch a section through matchboard wall panelling.

Activity

In an area designated by your trainer, fit flush dado panelling.

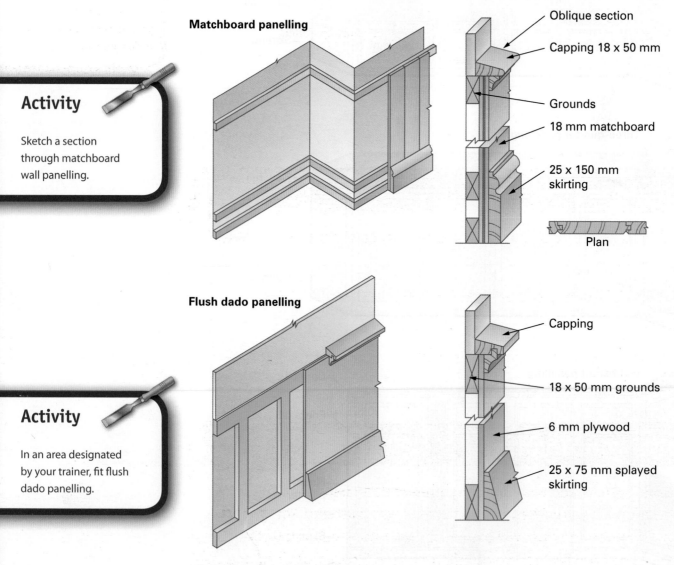

Matchboard panelling

Oblique section

Capping 18 x 50 mm

Grounds

18 mm matchboard

25 x 150 mm skirting

Plan

Flush dado panelling

Capping

18 x 50 mm grounds

6 mm plywood

25 x 75 mm splayed skirting

Figure 8.18 Three types of panelling (*continued opposite*)

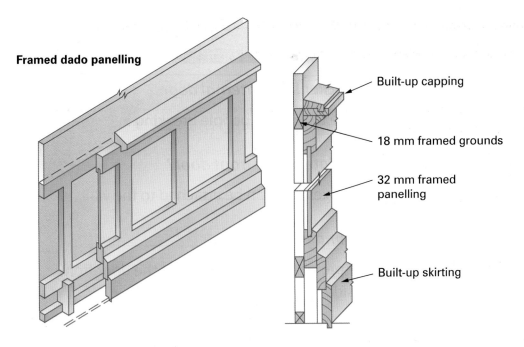

Framed dado panelling

Built-up capping

18 mm framed grounds

32 mm framed panelling

Built-up skirting

Figure 8.18 Three types of panelling (*continued*)

Activity

Sketch a section through framed dado panelling.

Activity

In an area designated by your trainer, fit framed dado panelling.

Knowledge refresher

1 Name five different types of moulding that can be fitted in a room.

2 Name three different profiles that can be used for mouldings.

3 Name three different types of wall panelling.

4 Give an example of where wall panelling can be found.

What would you do?

You have been tasked with installing framed dado panelling. The framed grounds were fitted before the wall was plastered. You go to fit the panelling but on one wall the grounds are not level, and on another they are uneven, with some ground protruding 5 mm more from the plaster than others. What has caused this? What can be done to rectify it? What could the cost implications be?

FAQ

Why are external doors thicker than internal doors?

They need to be more secure to prevent people breaking in.

Do I need to have a letter plate on my front door?

No. There are secure letter boxes that can be attached to the wall.

Do I have to use security hinges?

No, but security hinges are preferable as they prevent the door from being forced at the hinge side.

Why can't I have more than two deadlocks on a single exterior door?

Any more than two deadlocks would weaken the door at the lock side, as you will have removed a lot of material when mortising the locks in.

What type of door would I fit a security chain on?

Any door that doesn't have a glass panel or spy hole in it.

What is the best door closer to use?

The choice of door closer depends on the client's preference, but usually internal doors have concealed closers while doors in corridors have overhead closers.

Can I choose to use a dado profile for a picture rail?

If it is for decoration only, yes, but if it is to support paintings, no, because profiles for dado are not suitable for this.

chapter 9

Structural carcassing

OVERVIEW

Structural carcassing covers all carpentry work associated with the structural elements of a building such as floors and roofs. This chapter is designed to help you identify the main activities associated with structural carcassing and to provide you with the knowledge and understanding required to carry them out.

It will cover the following:

- basic terms – a refresher
- pitched roofs
- flat roofs
- ground and upper floors.

Basic terms – a refresher

Before looking at new and more complex areas of roofing, you should remind yourself of some of the terms and ideas you came across at Level 2. We will recap the key terms briefly now.

Roofing terminology

Roofs are made up of a number of different parts called 'elements'. These in turn are made up of 'members' or 'components'.

Elements

The main elements are defined below and shown in Figure 9.1.

- **gable** – the triangular part of the end wall of a building that has a pitched roof
- **hip** – where two external sloping surfaces meet
- **valley** – where two internal sloping surfaces meet
- **verge** – where the roof overhangs at the gable
- **eaves** – the lowest part of the roof surface where it meets the outside walls.

Members or components

The main members or components are defined below and shown in Figure 9.1.

- **ridge board** – a horizontal board at the apex acting as a spine, against which most of the rafters are fixed
- **wall plate** – a length of timber placed on top of the brickwork to spread the load of the roof through the outside walls and give a fixing point for the bottom of the rafters
- **rafter** – a piece of timber that forms the roof, of which there are several types
- **common rafters** – the main load-bearing timbers of the roof
- **hip rafters** – used where two sloping surfaces meet at an external angle, this provides a fixing for the jack rafters and transfers their load to the wall
- **crown rafter** – the centre rafter in a hip end that transfers the load to the wall
- **jack rafters** – these span from the wall plate to the hip rafter, enclosing the gaps between common and hip rafters, and crown and hip rafters
- **valley rafters** – like hip rafters but forming an internal angle, acting as a spine for fixing cripple rafters
- **cripple rafters** – similar to a jack rafter, these enclose the gap between the common and valley rafters

- **purlins** – horizontal beams that support the rafters mid-way between the ridge and wall plate.

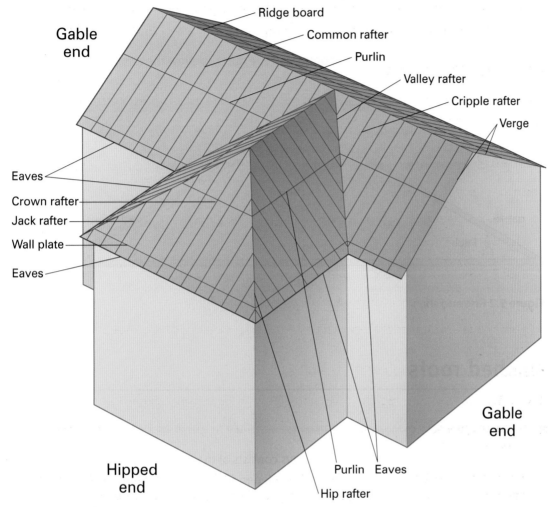

Figure 9.1 Roofing terminology

Basic terms for setting out

Setting out covers all the action necessary before commencing construction. Here are the most common terms you will need, which were covered in *Carpentry and Joinery NVQ and Technical Certificate Level 2 2nd edition* pages 331–332.

- **span** – the distance measured in the direction of the ceiling joists, from the outside of the wall plate to wall plate, known as the overall span

- **run** – equal to half the span

- **apex** – the peak or highest part of the roof

- **rise** – the distance from the outside of the wall plates at wall-plate level to the apex

- **pitch** – the angle or slope of the roof, calculated from the rise and the run

Did you know?

The span, run and rise are measured in metric units but the pitch is measured in degrees.

- **pitch line** – a line that is marked up from the underside of the rafter, one third of its depth to the top of the birdsmouth cut
- **plumb cut** – the angle of cut at the top of the rafter
- **seat cut** – the angle of cut at the bottom of the rafter
- **birdsmouth** – notch cut out at the bottom of the rafter to allow the rafter to sit on the wall plate.

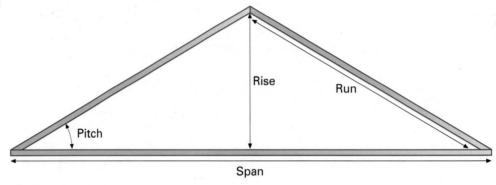

Figure 9.2 Rise and span

Pitched roofs

Trussed roofs

Most roofing on domestic dwellings these days consists of trusses, factory-made from stress-graded **PAR** timber to a wide variety of designs, depending on requirements. Using trussed roofing saves on timber and makes the process of roofing easier and quicker than with a traditional pitched roof. Trussed roofs can also span greater distances without the need for support from intermediate walls.

Definition

PAR – planed all round

In a trussed roof, each truss is composed of triangles, because of the structural stability of that shape. All joints are butt-jointed and held together with fixing plates, face-fixed on either side. These plates are usually made of galvanised steel, either nailed or factory-pressed; they can also be gang-nailed gusset plates made of 12 mm resin-bonded plywood.

Handling and storing truss rafters

Truss rafters should be stored on raised bearers to avoid contact with the ground, and should be supported as to prevent any distortion. Trusses should ideally be stored vertically on bearers located at the points of support, with suitable props to maintain them vertically. Alternatively, trusses can be stored horizontally, again on bearers and supported at each joint.

Manhandling trusses should be kept to a minimum due to the risk of manual handling injuries. When they have to be handled – for example, when they are transported – they must be handled with great care and supported correctly along any stress points.

Mechanical handling of trusses is usually done with a crane, when they should be lifted in banded sets and lowered onto suitable bearing supports. Suggested lifting points are at rafter or ceiling node points. A suitable spreader bar should be used to withstand any sling forces, and a guide rope should be attached to allow for precise movement.

Truss types

There are a variety of truss styles, the most common of which are shown below.

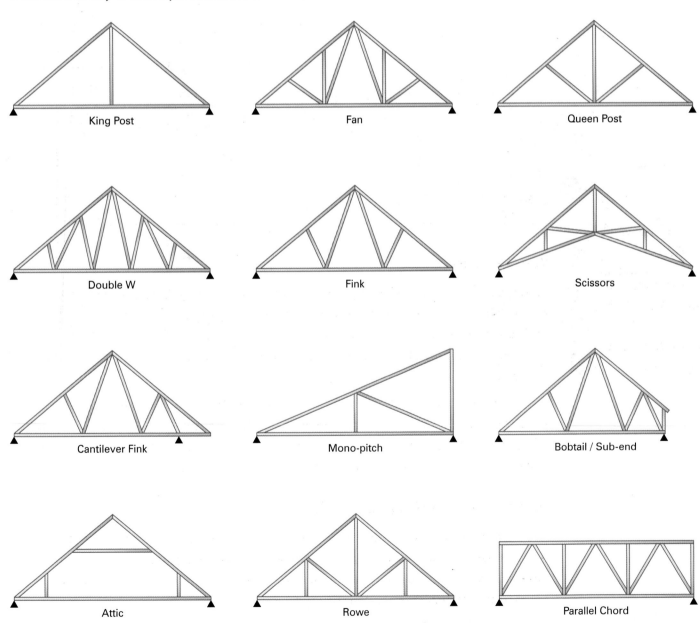

Figure 9.3 Caption

Erecting trusses

Wall plates are bedded onto the wall and then onto the wall plate. The position of the trusses can be marked out at a maximum of 600 mm centres, then truss clips can be used or the trusses can be cheek-nailed (skew nailed) into the wall plate.

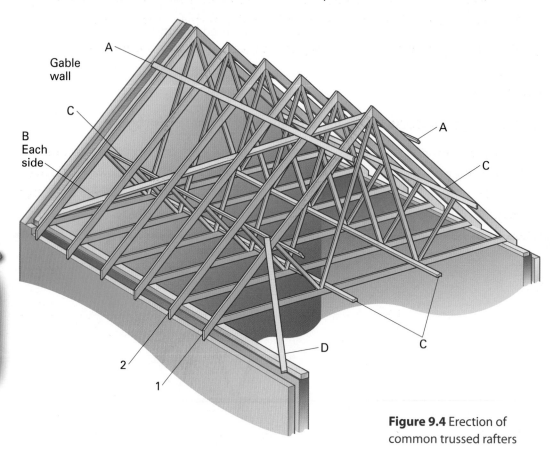

Activity

In an area designated by your trainer, mark out and fix a truss rafter roof.

Figure 9.4 Erection of common trussed rafters

Remember

Never alter a trussed rafter without the structural designer's approval.

Step 1 Fix first truss using framing anchors 1.

Step 2 Stabilise and plumb first truss with temporary braces E.

Step 3 Fix temporary battens on each side of ridge A.

Step 4 Position next truss 2.

Step 5 Fix the wall plate and temporary battens A. Continue until last truss is positioned.

Step 6 Fix braces B.

Step 7 Fix braces C.

Step 8 Fix horizontal restraint straps at max 2.0 m centres across trusses on to the inner leaf of the gable walls.

Fixing bracing

Fixing bracing is important for the safety and stability of any truss roof. There are two main types of bracing:

1   stability bracing – this holds the trusses firmly in place and keeps them straight, so that they can withstand all the loads on the roof, with the exception of wind loads

2   wind bracing – this is additional bracing that is used to withstand the wind loads.

Gable ends

Setting out for a gable end

First set out and fix the wall plate. The wall plate is set on the brick or block work and either bedded in by the bricklayer or temporarily fixed by nailing through the joints. Once secured it is held in place with **restraint straps** (see Figure 9.5). If the wall plate is to be joined in length, a **halving joint** is used (see Figure 9.6). It is vital that the wall plate is fixed level to avoid serious problems later.

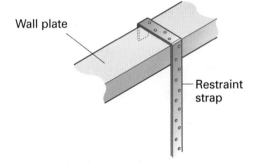

Figure 9.5 Restraint straps

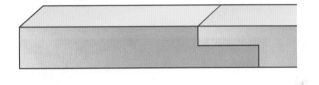

Figure 9.6 Plate with halving joint

Once the wall plate is in place, you need to measure the span and the rise. You can use these measurements to work out the rafter length in different ways, using a **roofing ready reckoner**, **geometry** or **scale drawings**. Ready reckoners and geometry are covered later in this chapter, so we will start with scale drawings.

For this example we will use a span of 5m and a rise of 2.3m.

Using a scrap piece of plywood or hardboard we first draw the roof to a scale that will fit the scrap piece of plywood/hardboard (usually a scale of 1:20).

From this drawing we can measure at scale and find the true length of the rafter. Then by using a **sliding bevel** we can work out the plumb cut and seat cut.

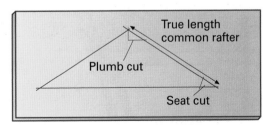

Figure 9.7 Sketch on a piece of scrap ply showing the true length, plumb and seat cut for a common rafter

Definition

Geometry – a form of mathematics using formulas, arithmetic and angles

Scale drawing – a drawing of a building or component in the right proportions, but scaled down to fit on a piece of paper. On a drawing at a scale of 1:50, a line 10 mm long would represent 500 mm on the actual object

Sliding bevel – a tool that can be set so that the user can mark out any angle

Making and using a pattern rafter

From our scale drawing we can mark out one rafter, which we will then use as a **pattern rafter**.

There are five easy steps to follow when marking out a pattern rafter.

Step 1 Mark the pitch line one-third of the way up the width of the rafter.

Step 2 Set the sliding bevel to the plumb cut and mark the angle onto the top of the rafter.

Step 3 Mark the true length on the rafter, measuring along the pitch line.

Step 4 Use the sliding bevel to mark out the seat cut, then with a combination square mark out the birdsmouth at 90 degrees to the seat cut.

Step 5 Re-mark the plumb cut to allow for half the thickness of the ridge.

Once it has all been marked out, this can be cut and used as a pattern rafter.

The pattern rafter can be used to mark out all the remaining common rafters, although it is advisable to mark out and cut only four, then place two at each end of the roof to check whether the roof is going to be level.

Once all the rafters are cut, mark out the wall plate and fix the rafters. Rafters are normally placed at 400 mm centres, with the first and last rafter 50 mm away from the gable wall. The rafters are usually fixed at the foot by skew-nailing into the wall plate and at the head by nailing through the ridge board.

Did you know?

The first and last rafters are placed 50 mm away from the wall to prevent moisture that penetrates the outside wall coming into contact with the rafters, thus preventing rot.

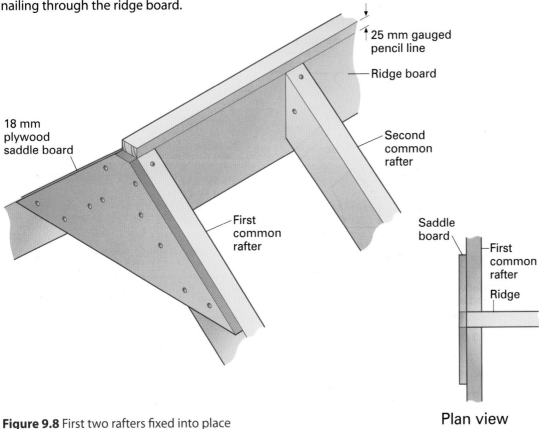

Plan view

Figure 9.8 First two rafters fixed into place

If the roof requires a loft space, joists can be put in place and bolted to the rafters; if additional support is required, struts can be used.

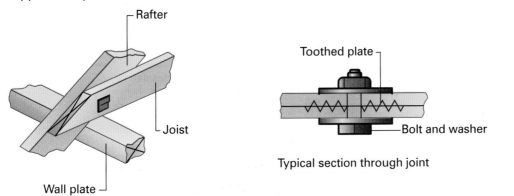

Typical section through joint

Figure 9.9 Rafter legs bolted to joists

For roofs with a large span, purlins provide adequate support. Purlins are usually built into the brickwork: either the gable wall will not have been finished being built yet or the bricklayer will have left a **sand course**. The same is true of roof ladders (see below).

Finishing a gable end

In Chapter 5, you saw how to use a roof ladder (page 131). Here we look at marking out and fixing a bargeboard.

The simplest way of marking out the bargeboard is to temporarily fix it in place and use a level to mark the plumb and seat cut.

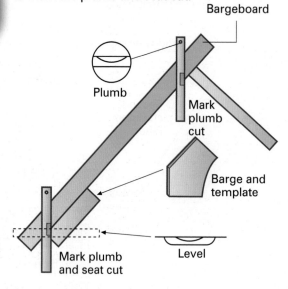

Figure 9.10 Marking out a bargeboard using a level

When fixing a bargeboard, the foot of the board may be mitred to the fascia, butted and finished flush with the fascia, or butted and extended slightly in front of the fascia to break the joint.

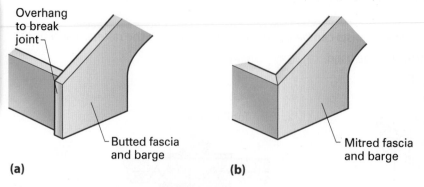

Figure 9.11 Fascia joined to bargeboard

The bargeboard should be fixed using oval nails or lost heads at least 2.5 times the thickness of the board so that a strong fixing is obtained. If there is to be a joint along the length of the bargeboard, the joint *must* be a mitre.

Activity

In an area designated by your trainer, mark out and fit a traditional cut roof with gable ends.

Hipped roofs

In Chapter 5, you learned about marking out for a hipped roof, and how to find true lengths, using Pythagoras' theorem. Here we look at the rest of the marking and setting out process.

From this point, to make the geometrical drawings as clear as possible, abbreviated labels will be used. See Table 9.1.

Abbreviation	Definition	Abbreviation	Definition
TL	true length	PCHR	plumb cut hip rafter
TLCR	true length common rafter	SCHR	seat cut hip rafter
TLHR	true length hip rafter	PCVR	plumb cut valley rafter
TLJR	true length jack rafter	SCVR	seat cut valley rafter
TLCrR	true length cripple rafter	EC	edge cut
PC	plumb cut	ECHR	edge cut hip rafter
SC	seat cut	ECJR	edge cut jack rafter
PCCR	plumb cut common rafter	ECVR	edge cut valley rafter
SCCR	seat cut common rafter	ECCrR	edge cut cripple rafter

Table 9.1 Abbreviations

There are two other angles that are concerned with hip rafters: the dihedral angle (or backing bevel for the hip) and the edge cut to the hip.

Finding the dihedral or backing bevel angle

The backing bevel angle is the angle between the two sloping roof surfaces. It provides a level surface so that the tile battens or roof boards can lie flat over the hip rafters. The backing bevel angle is rarely used in roofing today as the edge of the hip is usually worked square, but you should still know how to work it out.

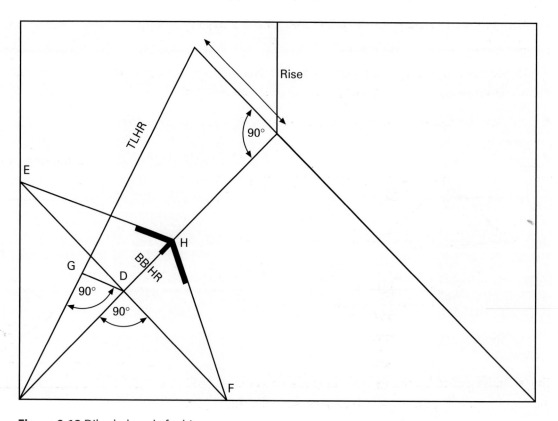

Figure 9.12 Dihedral angle for hip

1. Draw a plan of the roof and mark on the TLHR as before.

2. Draw a line at right angles to the hip on the plan at D, to touch the wall plates at E and F.

3. Draw a line at right angles to the TLHR at G, to touch point D.

4. With centre D and radius DG, draw an arc to touch the hip at H.

5. Join E to H and H to F. This gives the required backing bevel (BBHR).

Finding the edge cut

The edge cut is applied to both sides of the hip rafter at the plumb cut. It enables the hip to fit up to the ridge board between the crown and the common rafters.

1. Draw a plan of the roof and mark on the TLHR as before.

2. With centre I and radius IB, swing the TLHR down to J, making IJ, TLHR.

3. Draw lines at right angles from the ends of the hips and extend the ridge line. All three lines will intersect at K.

4. Join K to J. Angle IJK is the required edge cut (ECHR).

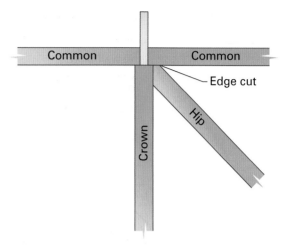

Figure 9.14 Edge cut on hip joined to ridge

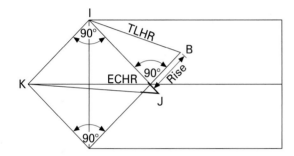

Figure 9.13 Edge cut on hip

The jack rafter's plumb and seat cuts are the same as those used in the common rafters so all you need to work out is the true length and edge cut.

- Draw the plan and section of the roof. Mark on the plan the jack rafters. Develop roof surfaces by swinging TLCR down to L and projecting down to M¹

- With centre N and radius NM¹, draw arc M¹O. Join points M¹ and O to ends of hips as shown.

- Continue jack rafters on to development.

- Mark the true length of jack rafter (TLJR) and edge cut for jack rafter (ECJR).

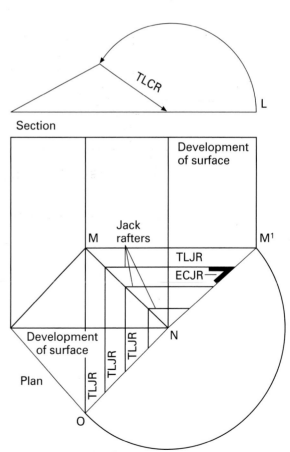

Figure 9.15 Jack rafter length and cuts

Setting out a hipped end

First we need to fit the wall plate, then the ridge and common rafters. To know where to place the common rafters you need to work out the true length of the rafter, then begin to mark out the wall plate.

The wall plate is joined at the corners and marked out as shown if Figure 9.28.

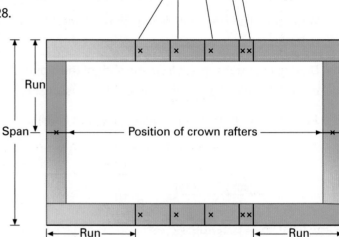

Figure 9.17 Wall plate marked out

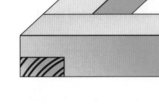

Figure 9.16 Corner halving

Marking out for a hipped roof

To mark out for a hipped roof, follow these steps:

1. Measure the span and divide it by two to get the run, then mark this on the hipped ends – the centre of your crown rafter will line up with this mark.

2. Mark the run along the two longer wall plates – these marks will give you the position of your first and last common rafter.

3. The common rafter will sit to the side of this line, so a cross or other mark should be made to let you know which side of the line the rafter will sit.

4. Mark positions for the rest of the rafters on the wall plate at the required centres, again using a cross or other mark to show you which side of the line the rafter will sit. Note: the last two rafters may be closer together than the required centres but must not be wider apart than the required centres.

5. Cut and fit the common rafters using the same method as used for a gable roof.

6. Fit the **crown rafter**, which has the same plumb and seat cuts as the common rafter and is almost the same length – but here you should *not* remove half the ridge thickness.

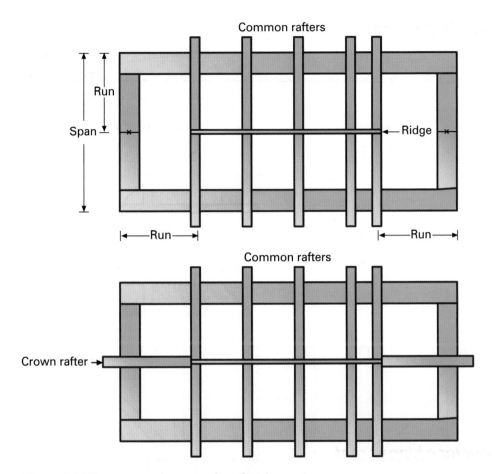

Figure 9.18 Common and crown rafters fitted

7. For the hip rafters, work out the true length, all the angles and bevels and mark out one hip as shown below, then cut the hip and try it in the four corners. If the hip fits in all four corners, you can use it as a template to mark out the rest of the hips; if not, the roof is out of square or level, but you can still use this hip to help mark out the remaining three corners.

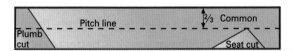

Figure 9.19 Pitch line marked on hip

With a hip rafter it is important to remember that the pitch line is marked out differently. It is marked from the top of the rafter and is set at two-thirds of the depth of the *common* rafter.

The best way to check your rafter before cutting is to measure from the point of the ridge down to the corner of the wall plate. This distance should be the same as marked on your rafter.

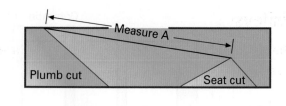

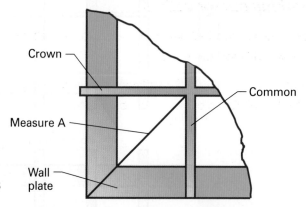

Figure 9.20 Checking hip measurements

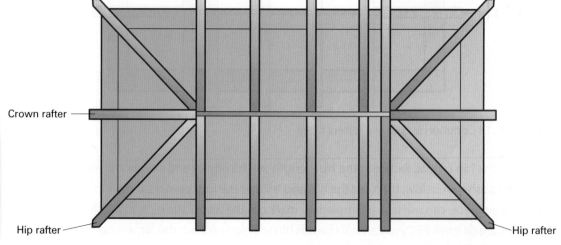

Figure 9.21 Hip rafters in place

8. Cut in the jack rafters. Find out the true length and edge cut of all the jack rafters, then mark them out and cut them. As the jack rafters are of different sizes it is better to cut them individually to fit. They can still be used as template rafters on the opposite side of the hip.

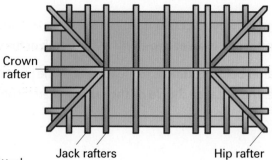

Figure 9.22 Jacks fitted

Valleys

The next section deals with valleys, which are formed when two sloping parts of a pitched roof meet at an internal corner.

Marking out for valleys

Valleys can be worked out in the same way as hips, using either a ready reckoner or geometry. Here we will look at geometry.

Working out the angles for valleys is similar to doing so for hips except that the key drawing is not a triangle but a plan drawing of the roof.

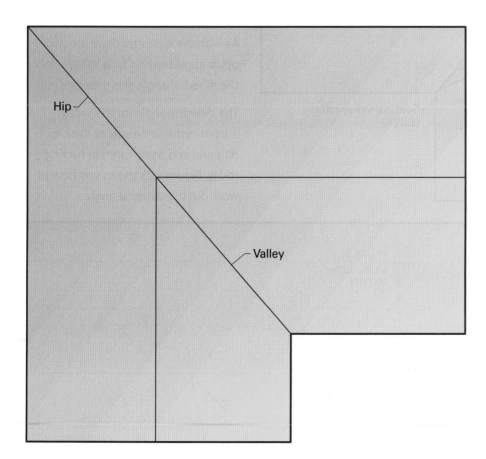

Figure 9.23 Plan of roof

First you need to find out the valley rafter true length, plumb and seat cut. Start by finding the rise of the roof and drawing a line this length at a right angle to the valley where it meets the ridge. Join this line to the point where the valley meets the wall plate. This will give you the true length of the valley rafter as well as the plumb and seat cuts.

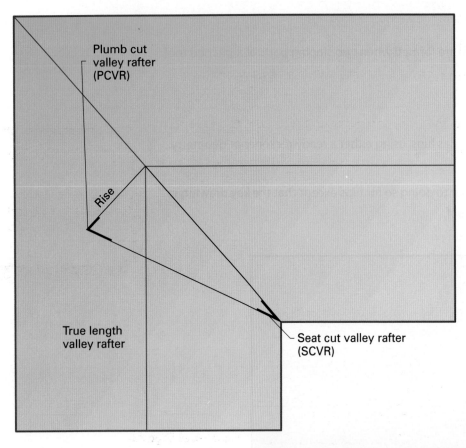

Figure 9.24 Valley true length

As with the hip rafter there are two other angles to find for a valley rafter: the dihedral angle and the edge cut.

The dihedral angle for the valley is used in the same way as the hip dihedral and again rarely in roofing today. Figure 9.25 shows you how to work out the dihedral angle.

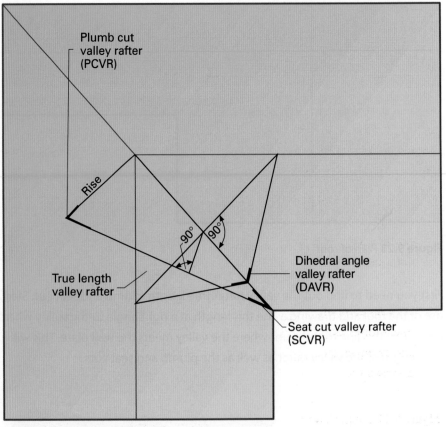

Figure 9.25 Dihedral angle hip

The final angle to find is the edge cut for the valley rafter, as follows.

1. Mark on the rise and true length of the valley rafter.

2. Draw a line at right angles to the valley where it meets the wall plate and extend this line to touch the ridge at A.

3. Set your compass to the true length of the valley and **swing an arc** towards the ridge at B.

4. Join up the line A–B to give you the edge cut.

Definition

Swing an arc – set a compass to the required radius and lightly draw a circle or arc

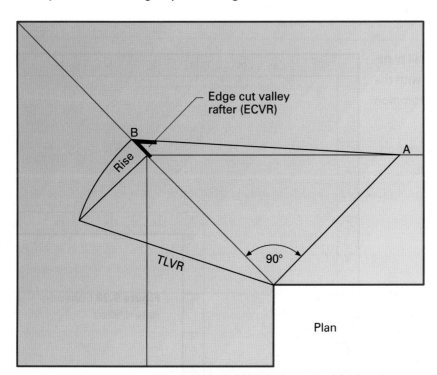

Figure 9.26 Edge cut

The final part of valley geometry is to find the true length and edge cut for the cripple rafters, as follows.

1. Draw out the roof plan as usual, then to the side of your plan draw out a section of the roof.

2. Set your compass to the rafter length and swing an arc downwards.

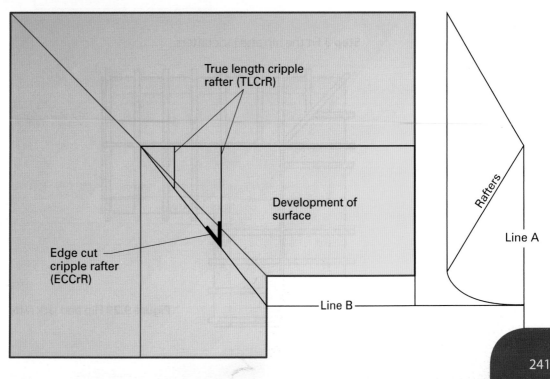

Figure 9.27 Cripple length and angle

3. Draw line A downwards until it meets the arc, then draw a line at right angles to line A until it hits the wall plate, creating line B.

4. Draw a line from where line B hits the wall plate up to where the valley meets the ridge. This will give you the appropriate true length (TLCrR) and edge cut (ECCrR).

Setting out a valley

There are four steps to follow when setting out a valley.

Step 1 Fit the wall plate and mark it out with the position of the common rafters.

Step 2 Fit the common rafters and ridge.

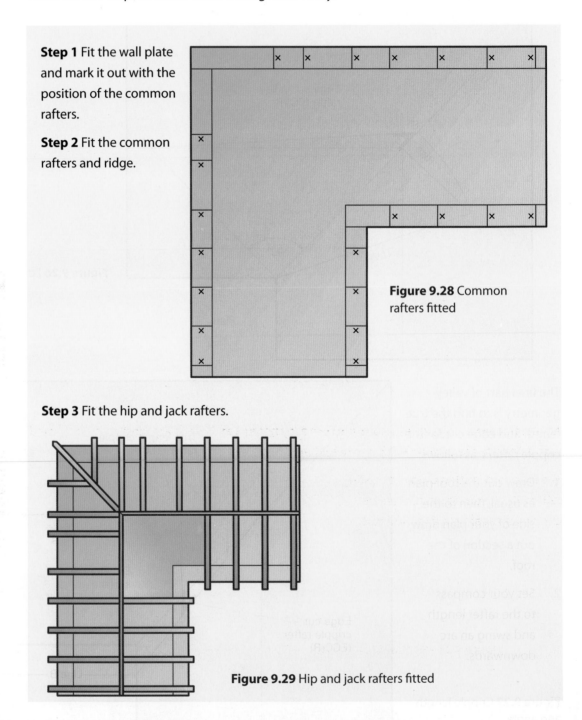

Figure 9.28 Common rafters fitted

Step 3 Fit the hip and jack rafters.

Figure 9.29 Hip and jack rafters fitted

Step 4 Fit the valley and cripple rafters, taking the true lengths and bevels from the drawings.

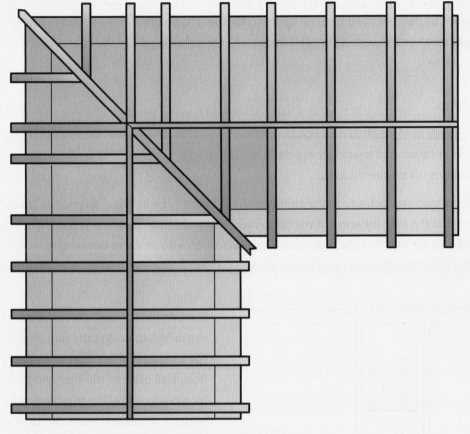

Figure 9.30 Valley and cripple rafters fitted

Activity

In an area designated by your trainer, mark out and fit a traditional cut roof with a valley.

An alternative to using valley rafters is to use a lay board. **Lay boards** are most commonly used with extensions to existing roofs or where there are dormer windows.

The lay board is fitted onto the existing rafters at the correct pitch, then the cripple rafters are cut and fixed to it.

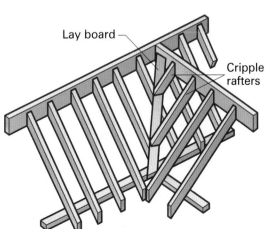

Lay board

Cripple rafters

Figure 9.31 Valley lay board set-up

Definition

Lay board – a piece of timber fitted to the common rafters of an existing roof to allow the cripple rafters to be fixed

Definition

Trimming an opening – removing structural timbers to allow a component such as a chimney or staircase to be fitted, and adding extra load-bearing timbers to spread the additional load

Trimming, eaves and covering

Trimming roof openings

Roofs often have components such as chimneys or roof windows. These components create extra work, as the roof must have an opening for them to be fitted. This involves cutting out parts of the rafter and putting in extra support to carry the weight of the roof over the missing rafters.

Chimneys

Chimneys are rarely used in new house construction as there are more efficient and environmental ways of heating these days, but most older houses will have chimneys and these roofs must be altered to suit.

When constructing such a roof, the chimney should already be in place, so you should cut and fit the rest of the roof, leaving out the rafters where the chimney is. When you mark out the wall plate, make sure that the rafters are positioned with a 50 mm gap between the chimney and the rafter. You may also need to put in extra rafters.

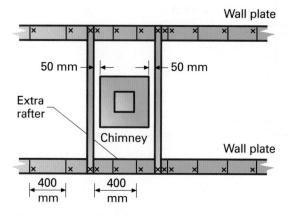

Figure 9.32 Wall plate marked out to allow for chimney

Next, fit the trimmer pieces between the rafters to bridge the gap, then fix the trimmed rafters – rafters running from wall plate to trimmer and from trimmer to ridge. The trimmed rafters are birdsmouthed at the bottom to sit over the wall plate, and the plumb and seat cut is the same as for common rafters.

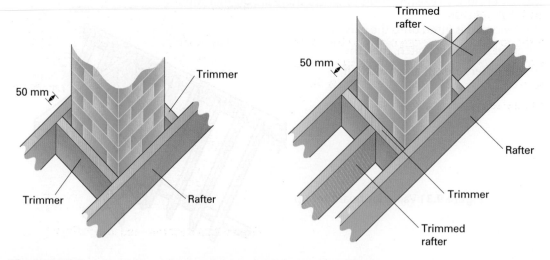

Figure 9.33 Opening trimmed around chimney with trimmers fitted

If the chimney is at **mid-pitch** rather than at the ridge, you will need to fit a chimney/back gutter: this ensures the roof remains watertight by preventing the water gathering at one point. The chimney gutter should be fixed at the back of the stack.

Definition

Mid-pitch – in the middle of the pitch rather than at the apex or eaves

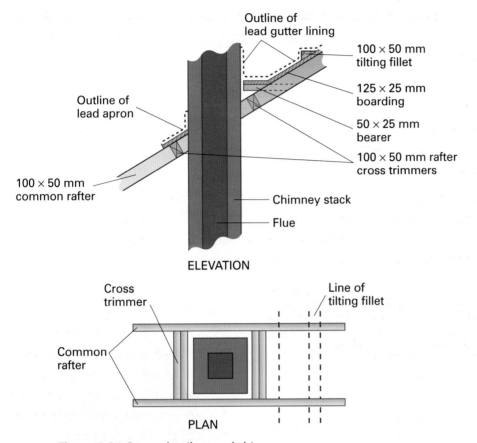

Figure 9.34 Gutter detail around chimney

Roof windows and skylights

You need to trim openings for roof windows and skylights too.

If the roof is new, you can plan it in the same way as for a chimney, remembering to make sure the area you leave for trimming is the same size as the window.

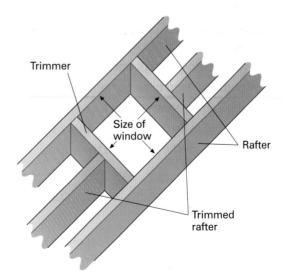

Figure 9.35 Roof trimmed for roof light

If you need to fit a roof window in an existing roof the procedure is different.

1. Strip all the tiles or slates from the area where you want to fit the window.

2. Remove the tile battens, felt and roof cladding.

3. Mark out the position of the window and cut the rafters to suit.

4. Now fit the trimmer and trimmed rafters– you may need to double up the rafters for additional strength – and fit the window.

5. Re-fit the cladding and felt, and fix any flashings.

6. Cut and fit new tile battens and finally re-tile or re-slate the roof.

Dormers

Dormer windows are different to roof windows: a roof window lays flat against the roof while a dormer window projects up from the sloping surface. A dormer window is often preferred to a roof window when there is limited headroom.

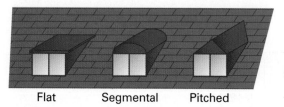

Flat   Segmental   Pitched

Figure 9.36 Dormer types

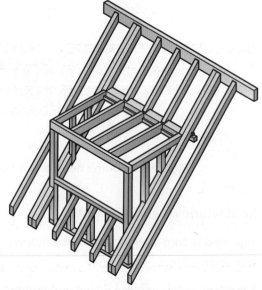

The construction of a dormer is similar to that of a roof window, except once the opening is trimmed, you need to add a framework to give the dormer shape. For a dormer you usually double up the rafters on either side to give support for the extra load the dormer puts on the roof.

Figure 9.37 Framing for dormer

Eaves details

The eaves are how the lower part of the roof is finished where it meets the wall, and incorporates fascia and soffit.

For more details about eaves, see Chapter 5, page 132.

Ready reckoner

A ready reckoner is a book used as an alternative to the geometry method and is often the simplest way of working out lengths and bevels. The book consists of a series of tables that are easy to follow once you understand the basics.

To use the ready reckoner you must know the span and the pitch of the roof.

Example

Take a hipped roof with a 36 degree pitch and a span of 8.46m. First you halve the span, getting a run of 4.23m. Referring to the tables in the ready reckoner, you can work out the lengths of the common rafter as follows:

RISE OF COMMON RAFTER 0.727 m PER METRE RUN   PITCH 36 degrees

BEVELS: COMMON RAFTER   SEAT 36
           COMMON RAFTER   PLUMB 54
           HIP OR VALLEY   SEAT 27
           HIP OR VALLEY   RIDGE 63
           JACK RAFTER   EDGE 39

JACK RAFTERS   333 mm CENTRES DECREASE   412 mm
               400 mm CENTRES DECREASE   494 mm
               500 mm CENTRES DECREASE   618 mm
               600 mm CENTRES DECREASE   742 mm

Run of Rafter	0.1	0.2	0.3	0.4	0.5	0.6	0.7	0.8	0.9	1.0
Length of Rafter	0.124	0.247	0.371	0.494	0.618	0.742	0.865	0.989	1.112	1.236
Length of Hip	0.159	0.318	0.477	0.636	0.795	0.954	1.113	1.272	1.431	1.590

You can see that the length of a rafter for a run of 0.1 m is 0.124 m, therefore for a run of 1 m, the rafter length will be 1.24 m – but you need to find the length of a rafter for a run of 4.23 m. This is how:

      1.00 m = 1.24 m

So: 4.00 m = 4.994 m

      0.20 m = 0.247 m

      0.03 m = 0.037 m

     total = 5.228 m

So the length of the common rafter is 5.228 m. However there are a few adjustments you must make before finding the finished size.

You need to allow for an overhang and for the ridge, which can both easily be measured. For the purposes of this example, we will use an overhang of 556 mm and a ridge of 50 mm. The final calculation is:

Basic rafter 5.228 m

+ overhang 0.556 m

– half ridge 0.025 m

total **5.759 m**

So the rafter length is 5.759 m.

Now you need to refer back to the table, which tells you that, for the common rafter, the seat cut is 36 degrees and the plumb cut is 54 degrees.

You can now mark out and cut your pattern rafter as before, remembering to mark on the pitch line and plumb cut. Measure the original size (5.228 m for our example) along the plumb cut, mark out the seat cut and finally cut.

The hip and valley rafters are worked out in the same way but jack and cripple rafters are different.

Jack and cripple rafters are marked on the table and are easy to work out. Continuing with the example, where the rafter length is 5.228 m, you look at the table and, if you are working at spacing the rafters at 400 mm centres, you reduce the length of the common rafter by 494 mm. The first jack/cripple rafter will be 5.228 m – 0.494 m = 4.734 m; the next jack/cripple rafter will be 4.734 m – 0.494 m = 4.240 m and so on. The angles for all the rest of the cuts are all shown on the table.

BEVELS:	COMMON RAFTER	SEAT 36
	COMMON RAFTER	PLUMB 54
	HIP OR VALLEY	SEAT 27
	HIP OR VALLEY	RIDGE 63
	JACK RAFTER	EDGE 39

JACK RAFTERS	333 mm CENTRES DECREASE	412 mm
	400 mm CENTRES DECREASE	494 mm
	500 mm CENTRES DECREASE	618 mm
	600 mm CENTRES DECREASE	742 mm

Remember

The plumb and seat cuts for the jack rafter are the same as for the common rafter.

Knowledge refresher

1. How should trussed rafters be stored?

2. How should trussed rafters be handled?

3. What types of bracing are required in a roof?

4. What distance must the first and last common rafters must be from the wall?

5. What is the dihedral angle (backing bevel) and what purpose does it serve?

6. Why do you not need to work out the plumb and seat cuts for the jack and cripple rafters?

7. Describe two different ways of forming a valley.

8. Why do you need to add extra rafters when trimming an opening?

What would you do?

You have been tasked with cutting in a garage roof with gable ends. You work out the angles and lengths, and cut the pattern rafter, which you use to mark out the other rafters. You cut and fit the rafters and are almost finished when you decide to put a level on the ridge. You find that the roof is out of level. What could have caused this? What could have been done to prevent it?

Construction of a flat roof

Flat roof joists are similar in construction to floor joists (discussed later in this chapter) but unless they are to be accessible, they are not so heavily loaded. Joists are, therefore, of a smaller dimension than those used in flooring.

There are many ways to provide a fall on a flat roof. The method you choose depends on what the direction of fall is and where on a building the roof situated.

Laying joists to a fall

This method is by far the easiest: all you have to do is ensure that the wall plate fixed to the wall is higher than the wall plate on the opposite wall or vice versa. The problem is that this method will also give the interior of the roof a sloping ceiling. This may be fine for a room such as a garage, but for a room such as a kitchen extension the client might not want the ceiling to be sloped and another method would have to be used.

Joists with firring pieces

Firring pieces provide a fall without disrupting the interior of the room, but involve more work. Firrings can be laid in two different ways, depending on the layout of the joists and the fall.

The lay out of joists is explained in more detail in the flooring section of this chapter.

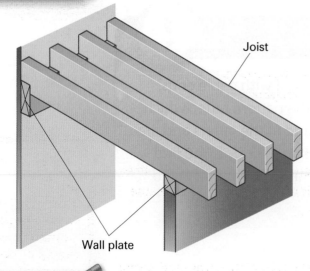

Joist

Wall plate

Figure 9.38 Joists laid to a fall

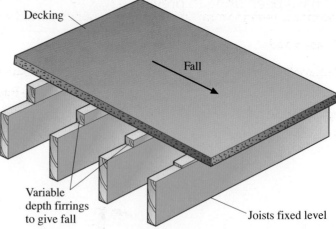

Decking

Fall

Variable depth firrings to give fall

Joists fixed level

Figure 9.39 Joists with firring pieces

Using firring pieces

The basic construction of a flat roof with firrings begins with the building of the exterior walls. Once the walls are in place at the correct height and level, the carpenter fits the wall plate on the eaves wall (there is no need for the wall plate to be fitted to the verge walls). This can be bedded down with cement, or nailed through the joints in the brick or block work with restraining straps fitted for extra strength. The carpenter then fixes the **header** to the existing wall.

The header can either be the same depth as the joists or have a smaller depth. If it is the same depth, the whole of the joist butts up to the header and the joists are fixed using joist hangers; if the header has a smaller depth, the joists can be notched to sit on top of the header as well as using framing anchors.

Once the wall plate and header are fixed, they are marked out for the joists at the specified centres (300 mm, 400 mm or 600 mm). The joists are cut to length, checked for **camber (crown)** and fixed in place using joist hangers. Strutting or noggings are then fitted to help strengthen the joists.

Once the joists are fixed in place they must have restraining straps fitted. A strap must be fitted to a minimum of one joist per 2m run, then firmly anchored to the wall to prevent movement in the roof under pressure from high winds.

The next step is to fix the firring pieces, which are either nailed or screwed down onto the top of the joists. Insulation is fitted between the joists, along with a vapour barrier to prevent the movement of moisture caused by condensation.

Decking

Once the insulation and vapour barrier are fitted, it is time to fit the decking. As you may remember from Level 2, decking a flat roof can be done with a range of materials including:

- **tongued and grooved board**

These boards are usually made from pine and are not very moisture-resistant, even when treated, so they are rarely used for decking these days. If used, the boards should be laid either with, or diagonal to, the fall of the roof. Cupping of the boards laid across the fall could cause the roof covering to form hollows in which puddles could form.

- **plywood**

Only roofing grade boards stamped with **WBP** (weather/water boil proof) should be used. Boards must be supported and securely fixed on all edges in case there is any distortion, which could rip or tear the felt covering. A 1 mm gap must be left between each board along all edges in case there is any movement caused by moisture, which again could cause damage to the felt.

Definition

Header – a piece of timber bolted to the wall to carry the weight of the joists

Camber (crown) – the bow or round that occurs along some timbers depth

Activity

In an area designated by your trainer, fit a flat roof using firring pieces.

Safety tip

Once the roof has been decked and is safe to walk on, the area around the roof (the verges and eaves) must be cordoned off with a suitable edge protection containing handrails and toeboards.

- **chipboard**

Only the types with the required water resistance classified for this purpose must be used. Boards are available that have a covering of bituminous felt bonded to one surface, giving temporary protection against wet weather. Once laid, the edges and joints can be sealed. Edge support, laying and fixing are similar to floors (covered later in this chapter). Moisture movement will be greater than with plywood as chipboard is more **porous**, so a 2 mm gap should be allowed along all joints, with at least a 10 mm gap around the roof edges. Tongued and grooved chipboard sheets should be fitted as per the manufacturer's instructions.

- **oriented strand board (OSB)**

Generally more stable than chipboard, but again only roofing grades must be used. Provision for moisture movement should be made as with chipboard.

- **cement-bonded chipboard**

Strong and durable with high density (much heavier and greater moisture resistance than standard chipboard). Provisions for moisture movement should be the same as chipboard.

- **metal decking**

Profiled sheets of aluminium or galvanised steel with a variety of factory-applied colour coatings and co-ordinated fixings are available. Metal decking is more usually associated with large steel sub-structures and fixed by specialist installers, but it can be used on small roof spans to some effect. Sheets can be rolled to different profiles and cut to any reasonable length to suit individual requirements.

- **translucent sheeting**

This might be corrugated or flat (e.g. polycarbonate twin wall), and must be installed as per the manufacturer's instructions.

Metal decking and translucent sheeting are supplied as finished products but timber-based decking needs additional work to it to make it watertight, as explained in the next section.

Weatherproofing a flat roof

Once a flat roof has been constructed and decked the next step is to make it watertight. The roof decking material can be covered using different methods and with different materials. One basic way of covering the decking is as follows.

The roof decking is covered in a layer of hot **bitumen** to seal any gaps in the joints. Then another layer of bitumen is poured over the top and felt is rolled onto it, sticking fast when the bitumen sets. A second roll of felt is stuck down with bitumen, but is laid at 90 degrees to the first. Some people add more felt at this stage – sometimes up to five more layers.

The final step can also be done in a number of ways. A cold system is now available, which replaces the need to use hot melt pots and blowtorches. Some people put stones or chippings down on top of the final layer; some use felt that has stones or chippings imbedded and a layer of dried bitumen on the back. The felt with stones imbedded into it is laid by rolling the felt out and using a gas blowtorch to heat the back, which softens the bitumen allowing it to stick.

Finishing a flat roof at the abutment and verge
Abutment

The abutment finish needs to take into consideration the existing wall as well as the flat roof. The abutment is finished by cutting a slot into the brick or block work and fixing lead to give a waterproof seal, which prevents water running down the face of the wall and into the room. **Tilt** or **angle fillets** are used to help with the run of the water and to give a less severe angle for the lead to be dressed to.

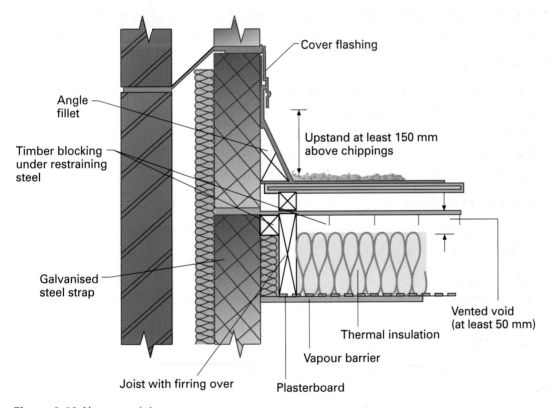

Figure 9.40 Abutment joint

Verge

Since a flat roof has such a shallow pitch, the verge needs some form of upstand to stop the water flowing over the sides at the verge instead of into the guttering at the eaves. This is done using tilt or angle fillets nailed to the decking down the full length of the verge prior to the roof being felted or finished.

Safety tip

Bitumen in liquid form is very hot, and if it comes in contact with the skin will stick and cause severe burns. When working with bitumen you must wear gloves and goggles, and ensure that arms and legs are fully covered.

Did you know?

Flat roofs have stones or chippings placed on top to reflect the heat from the sun, which can melt the bitumen, causing problems with leaks.

Definition

Tilt or angle fillet – a triangular-shaped piece of wood used in flat roof construction

Activity

Use a sketch to show how a flat roof is finished at an abutment.

Eaves details

The eaves details can be finished in the same way as the eaves on a pitched roof, with the soffit fitted to the underside of the joists and the fascia fitted to the ends.

Ground and upper floors

Several types of flooring are used in the construction of buildings, ranging from timber floors to large pre-cast concrete floors, which are used in large buildings such as residential flats. A suspended timber floor can be fitted at any level from top floor to ground floor. Details of flooring can be found in Chapter 5.

FAQ

Which is the best method to use when working out roof bevels and lengths?

There is no single best way: it all depends on the individual and what they find the easiest method to use.

When it comes to covering the roof, which method should I use?

The type of roof covering to be used depends on what the client and architect want, and on what is needed to meet planning and Building Regulations.

Why does there have to be a 50 mm gap between the rafters and the chimney?

Building Regulations state that there must be a 50 mm gap to prevent the heat from the chimney combusting the timber.

Why are flat roofs only guaranteed for a certain amount of time?

Most things that you buy or have fitted have a guarantee for a certain amount of time and building work is no different. If the flat roof had a guarantee for 50 years, the builder would be responsible for any maintenance work on the roof free of charge for 50 years. Since the average life of a flat roof is 12–15 years, the builder will only offer a guarantee for 10 years.

FAQ

Which type of flat roof decking is the best?

There is no specific best or worst but some materials are better than others. All the materials stated serve a purpose, but only if they are finished correctly, e.g. a chipboard-covered roof that is poorly felted will leak, as will a metal roof if the screw or bolt holes are not sealed correctly.

How can I get the lead on a flat roof to fit into the brick/block work?

The lead is fitted into a channel or groove that is cut into the brick/block work by a bricklayer. The groove is cut using either a disc cutter or a hammer and cold chisel. Once the groove is cut the lead is fed into the groove, wedged and sealed with a suitable mastic or silicon.

I have laid chipboard flooring and the floor is squeaking. What causes this and how can I stop it?

The squeaking is caused by the floorboards rubbing against the nails – something that happens after a while as the nails eventually work themselves loose. The best way to prevent this is to put a few screws into the floor to prevent movement – but be cautious: there may be wires or pipes under the floor.

Knowledge refresher

1   State two different ways of providing a fall on a flat roof.

2   Name four different materials used when decking a flat roof.

3   What is bitumen and how is it used?

What would you do?

You are tasked with fitting a flat roof. What method would you use? What type of decking would you use? What reasons do you have for your choices?

chapter 10

Maintaining components

OVERVIEW

In time, all components within a building will deteriorate. Timber will rot, brickwork crumble and even the ironmongery fail through wear and tear. These components need to be maintained or repaired to prevent the building falling into a state of disrepair. General building maintenance is now a recognised qualification, but traditionally this work is done by carpenters, as carpenters have the best links to other trades and the best understanding of what they do.

The maintenance of a building can range from periodically repainting the woodwork to replacing broken windows.

This chapter will not cover simpler remedial tasks such as replacing a door handle, but will cover the following basic maintenance tasks:

- timber decay

- replacing broken glazing

- repairing and replacing mouldings

- repairing a door that is binding

- repairing damaged windowsills and doorframes

- replacing window sash cords

- repairing structural timbers

- minor repairs to plaster and brickwork.

Timber decay

The decay of timber is caused by one or both of the following:

- wood-destroying fungi
- wood-boring insects.

Wood-destroying fungi

There many forms of fungi which, under suitable conditions, will attack timber until it is destroyed. They mainly fall into two groups:

1. dry rot
2. wet rot.

Both dry and wet rot can be a serious hazard to constructional timbers.

Dry rot

This is the most common and most serious of wood-destroying fungi. The fungus does most damage to softwoods but also attacks hardwoods, particularly when they are close to softwoods already infected. If left undetected or untreated it can destroy much of the timber in a building.

Dry rot is so called because of the dry, crumbly appearance of the infected timber. It is, however, excessive moisture that is the main cause of the decay. If the wood is kept dry and well ventilated there should be little chance of dry rot occurring.

The main conditions for an attack of dry rot are:

- damp timber, with a moisture content above 20 per cent (known as the **dry rot safety line**)
- poor, or no, ventilation (i.e. no circulation of fresh air).

The attack of dry rot takes place in three stages, which are shown in Figures 10.1 to 10.3:

- The spores (seeds) of the fungus germinate and send out hyphae (roots) which bore into the timber.
- The hyphae branch out and spread through the timber. A fruiting body now starts to grow.
- The fruiting body, which resembles a large, fleshy pancake, starts to ripen. When fully ripened it discharges millions of red spores into the air. The spores attach to fresh timber and the cycle starts again.

Definition

Dry rot safety line – when the moisture content of damp timber reaches 20 per cent. Dry rot is likely if the moisture content exceeds this

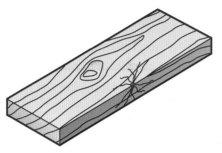

Figure 10.1 Dry rot stage one

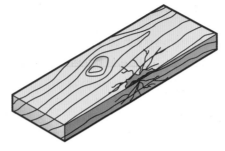

Figure 10.2 Dry rot stage two

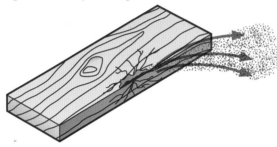

Figure 10.3 Dry rot stage three

Identification of dry rot

Dry rot can be identified by:

- an unpleasant, musty smell
- visible distortion of infected timber; warped, sunken and/or shrinkage cracks
- probing to test the timber for softening or crumbling
- the appearance of fruiting bodies
- presence of fine, orange-red dust on the floorboards and other parts of the structure
- presence of whitish-grey strands on the surface of the timber.

Eradication of dry rot

Dry rot is eradicated by carrying out the following actions:

- Eliminate all possible sources of dampness, such as blocked air bricks, bridged damp proof course, leaking pipes etc.
- Determine the extent of the attack.
- Remove all infected timber.
- Clean and treat surrounding walls, floors etc. with a suitable fungicide.
- Treat any remaining timber with a preservative.
- Replace rotted timber with new treated timber.
- Monitor completed work for signs of further attack.

Did you know?

Fungal fruiting bodies are the sign of advanced decay in timber.

Activity

In an area designated by your trainer, demonstrate the correct way to eradicate dry rot.

Remember

Timbers affected by dry rot, wet rot or insect attack should be cut back at least 600 mm.

Wet rot

This is a general name given to another type of wood-destroying fungus. The conditions where wet rot is found are usually wet rather than damp. Although wet rot is capable of destroying timber it is not as serious a problem as dry rot and, if the source of wetness is found, the wet rot can be halted.

The most likely places to find wet rot are:

- badly maintained external joinery, where water has penetrated
- ends of rafters and floor joists
- fences and gate posts
- under leaking sinks or baths etc.

Identification of wet rot

The signs to look for to identify wet rot are:

- timber becomes darker in colour with cracks along the grain
- decay usually occurs internally leaving a thin layer of relatively sound timber on the outside
- localised areas of decay close to wetness
- a musty, damp smell.

Eradication of wet rot

As wet rot is not as serious a problem as dry rot, less extreme measures are normally involved. It is usually sufficient to remove the rotted timber, treat the remaining timber with a fungicide and replace any rotted timber with treated timber. Lastly, if possible, the source of any wetness should be rectified.

Wood-boring insects

Many insects attack or eat wood, causing structural damage. There are four main types of insects found in the UK, all classed as beetles.

Common Furniture Beetle

This wood-boring insect can damage both softwoods and hardwoods. The larvae of the beetle bore through the wood, digesting the cellulose. After about three years, the beetle forms a pupal chamber near the surface, where it changes into an adult beetle. In the summer, the beetle bites its way out to the surface, forming the characteristic round flight hole, measuring about 1.5 mm in diameter. After mating, the females lay their eggs (up to 80) in cracks, crevices or old flight-holes. The eggs hatch and a new generation begins a fresh life cycle.

Death Watch Beetle

This wood-boring insect is related to the Common Furniture Beetle, but is much larger, with a flight hole of about 3 mm in diameter, usually found in decaying oak. The female lays up to 200 eggs. While generally attacking hardwoods only, this wood-boring insect has been known to feed on decaying softwood timbers. The Death Watch Beetle is well known for making a tapping sound, caused by the head of the male during the mating season.

Powder Post Beetle

This beetle gets its name from the way it can reduce timber to a fine powder. Powder Post beetles generally attack timber with a high moisture content. As with all other beetles, the female lays eggs and the larvae do the damage.

House Longhorn Beetle

This wood-destroying insect attacks seasoned softwoods, laying its eggs in the cracks and crevices of wood. In Great Britain, this insect is found mainly in Surrey and Hampshire.

As well as beetles, weevils and wood wasps can also cause a problem.

Activity

Using a sketch, show the lifecycle of a wood-boring insect.

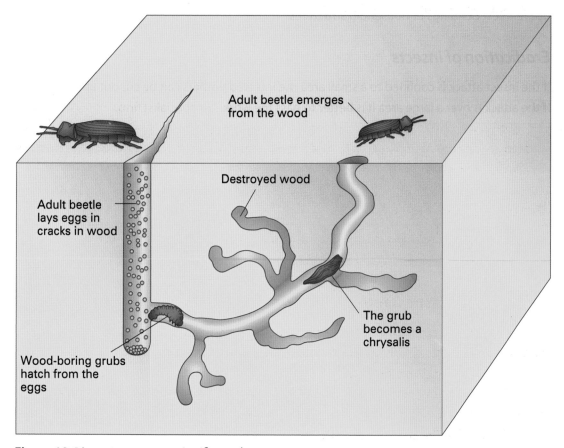

Adult beetle emerges from the wood

Destroyed wood

Adult beetle lays eggs in cracks in wood

The grub becomes a chrysalis

Wood-boring grubs hatch from the eggs

Figure 10.4 Insects can cause significant damage

Wood-boring Weevil

This is a wood-boring insect similar in appearance and size to the Common Furniture Beetle. It differs in that it will only attack timber that is already decayed by wood-rotting fungi. There are over 50,000 species of weevil, all of which have long snouts. The Wood-boring Weevil is prolific, and is known to have up to two complete life cycles in one year. Its presence may therefore be accompanied by serious structural collapse of timber.

Wood Wasp

These insects are also known as horntails. They are wasp-like insects and may be seen or heard buzzing and causing cosmetic damage around the home. The body is 2–5 cm long, and the wasp is multi-coloured, blue or black, with wings of matching colour. Walls, wood floors, doors and other wooden surfaces may have holes that are approximately 3 mm in diameter. These holes may also appear on items covering wooden structures, such as carpeting, wallpaper, and linoleum. Wood Wasps lay their eggs in forest trees that are either damaged or weak. It can take up to five years for adult Wood Wasps to emerge, so often they are found in trees that have been felled for construction purposes – often even after the wood has been stored. However, most Wood Wasp holes will appear within the first two years after use of the cut wood. The good news is that these insects only lay their eggs in weakened forest trees, and will not re-infest buildings or household structures.

Eradication of insects

If the insect attack is confined to a small area the infected timbers can be cut out and replaced. If the attack is over a large area it is better to pass the work to a specialist firm.

Remember

Contact a professional for advice where load bearing timbers have been attacked, as there may be a need to prop or shore up to prevent collapse of the structure.

Preservation and protection

Not only do we need to protect timber from fungal and insect attack but also, where timber is exposed, it needs protecting from the weather. Preservation extends the life of timber and greatly reduces the cost of maintenance.

Types of preservative

Timber preservatives are divided into three groups:

- tar oils
- water-borne
- organic solvents.

Tar oils

Tar oils are derived from coal tar, are very effective and relatively cheap. However, they give off a very strong odour which may contaminate other materials.

Water-borne

Water-borne preservatives are mainly solutions of copper, zinc, mercury or chrome. Water is used to carry the chemical into the timber and then allowed to evaporate, leaving the chemical in the timber. They are very effective against fungi and insects and are able to penetrate into the timber. They are also easily painted over and relatively inexpensive.

Organic solvents

Organic solvents are the most effective, but also the most costly of the preservatives. They have excellent penetrating qualities and dry out rapidly. Many of this type are proprietary brands such as those manufactured by Cuprinol™.

Methods of application

Preservative can be applied in two ways:

1. non-pressure methods – brushing, spraying, dipping or steeping
2. pressure methods – empty cell, full cell or double vacuum.

Non-pressure methods

Although satisfactory results can be achieved, there are disadvantages with using non-pressure methods. The depth of penetration is uneven and, with certain timbers, impregnation is insufficient to prevent leaking out (leaching). Table 10.1 lists non-pressure methods and how and where to employ them.

Method	How and where to employ it
Brushing	The most commonly used method of applying preservative, it is important to apply the preservative liberally and allow it to soak in
Spraying	Usually used where brushing is difficult to carry out, in areas such as roof spaces
Dipping	Timbers are submerged in a bath of preservative for up to 15 minutes
Steeping	Similar to dipping only the timber is left submerged for up to two weeks

Table 10.1 Non-pressure methods

Pressure methods

Pressure methods generally give better results with deeper penetration and less leaching. Table 10.2 lists pressure methods and how and where to employ them.

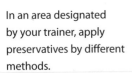

Activity

In an area designated by your trainer, apply preservatives by different methods.

Method	How and where to employ it
Empty cell	Preservative is forced into the timber under pressure. When the pressure is released the air within the cells expands and blows out the surplus for re-use. This method is suitable for water-borne and organic solvent preservatives.
Full cell	Similar to empty cell, but prior to impregnation a vacuum is applied to the timber. The preservative is then introduced under pressure to fill the cells completely. Suitable for tar oils and water-borne preservatives.
Double vacuum	A vacuum is applied to remove air from the cells, the preservative introduced, the vacuum released and pressure applied. The pressure is released and a second vacuum applied to recover surplus preservative. This method is used for organic solvent preservatives.

Table 10.2 Pressure methods

Knowledge refresher

1 How can dry rot be eradicated?

2 Name four types of insect that attack timber.

3 How can infestation by wood-boring insects be eradicated?

4 Name three non-pressure methods of applying preservative.

5 Describe how preservative is applied by double vacuum.

What would you do?

You have been tasked with re-insulating an attic space. When you remove the old insulation, you notice that the joists show signs of insect infestation. You inform the client, who thinks you are just trying to get extra work. How can you prove to the client that the infestation is real? What needs to be done to correct the infestation? What could the consequences be if the client ignores it?

Replacing broken glazing

This is a job done largely by specialist glazing companies, but there is still a call for the carpenter to replace broken glazing. This section will look at replacing both single and double glazed windowpanes.

Single glazing

Single glazing is mainly found in older buildings such as factories or country houses. The glass is fixed using putty and held in place with either putty tooled to a smooth finish or timber beading.

The first thing to do when replacing a pane of glass is to remove the old pane. If the glass is smashed, you must remove the remaining pieces very carefully, wearing suitable gloves and other necessary PPE. If the glass is only cracked, it is safer to tape the glass to prevent it shattering, then tap it gently from the inside until it falls out.

Next you need to clean out the rebate, removing all the old putty so that the new pane of glass can be fitted properly. Use a sharp chisel, taking care not to damage the rebate.

Now fit the new pane of glass. There are several methods for doing this: use the method that matches the existing glazing in the building.

Method 1 Putty only

This is the traditional method, best done with a putty knife.

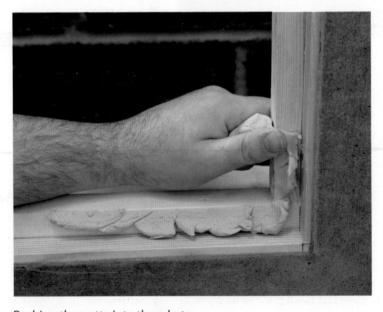

First make the putty useable – when it is first taken out of the tub, putty is very sticky and oily, and must be kneaded like dough to make it workable. Then push the putty into the rebate, taking a small amount of putty in your hand and feeding it into the rebate using your forefinger and thumb.

Once the putty is all around the rebate, offer the glass into place starting at the bottom, then apply a little pressure to ensure that the glass is squashed up against the putty. Do not apply too much pressure, as this will break the glass.

Pushing the putty into the rebate

Now insert glazing sprigs or panel pins to keep the glass in place until the putty sets. To avoid breaking the glass, place the edge of the hammer against the glass and slide it along the glass when driving in the pins or sprigs. If the hammer stays in contact with the glass, breakages are less likely. Drive the pins or sprigs in far enough that they are not in view, but not so far that they catch the edge of the glass and break it. Small panes need only one pin or sprig; larger panes will need more.

Take another ball of putty and feed it around the frame. Take the putty knife and, starting at the top, draw the knife down the window, squeezing the putty and shaping it at the same time. Repeat several times if necessary to get the desired finish. Remove any excess putty, including any on the inside that has been squeezed out from the rebate when the glass was pushed into place.

Sprigs being fitted

Putty being tooled

Method 2 Putty and timber beading

This is similar to Method 1, following the same steps until the glass has been pushed into place. At this point, instead of the putty being tooled along the outside, a timber bead is used.

First place putty on the face of the bead that will be pushed against the glass, again using your forefinger and thumb.

Activity

In an area designated by your trainer, replace a single pane of glass using a 'putty only' method.

Did you know?

When using a putty knife, it can be useful to dip the knife into water to prevent the putty sticking to the blade and dragging the putty.

Nailing the bead in place

Next fit the beads, which are usually secured using panel pins or small finish nails such as brads or lost heads. The beads should be fitted in sequence: top bead first, then the bottom and finally the sides.

Press the beads into place with enough force to squeeze the putty out, leaving just a small amount, then drive the pins or nails in as in Method 1. It can be difficult to put pressure on the bead and nail it at the same time, so ideally you need two people. Alternatively you can drive the nail into the bead a little way, but not far enough to penetrate the other side, then push the bead into place with one hand and drive the nail home with the other.

Once all the beads are fitted, remove the excess putty from both the inside and outside of the glass.

Method 3 Silicone and beads

This is identical to Method 2 but uses silicone instead of putty. It is often preferred as it is quicker and not as messy.

With advances in technology there are other ways of glazing. Rubber bead, tape or strips can be used to glaze single panes, but is more commonly used on double glazing. We will cover this method in the next section.

Double glazing

Double glazing can be found in either timber or uPVC windows. With uPVC double glazing, there are many different ways to install the glass using different types of beads and rubber components, so it is best to refer to the manufacturer when changing a unit. Here we will concentrate on replacing double-glazed units on timber windows.

When replacing double glazing, silicone and putty are rarely used, as a pre-made rubber bead, tape or strip is preferred.

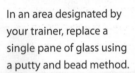

Activity

In an area designated by your trainer, replace a single pane of glass using a putty and bead method.

Remember

With silicone, take care not to put too much on: once the glass or the timber bead is pushed into place, the silicone spreads more than putty but does not clean off glass as easily.

Types of rubber bead, tape and strips

Various types of bead, tape and strip are available. Just a few examples are shown in Figure 10.5.

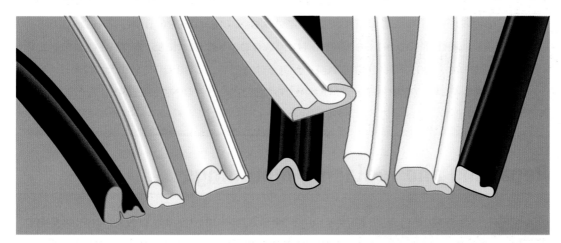

Figure 10.5 Some of the rubber beads, tape and strips available

First you need to remove the damaged or broken unit. As with single glazing, if the unit is just cracked, use tape to stop it shattering. Then remove the timber beads that keep the window in place, taking care not to damage them or the rubber if they are to be used again.

Now, wearing the correct protective clothing, carefully take out the glass. If the glass was installed with double-sided sticky rubber tape, run a sharp utility knife around the inside to free the glass. Next clean the beads and the rebate and, if you are using double-sided sticky rubber tape, apply this to the rebate.

Now the new double-glazed unit can be fitted and the beads pinned back into place.

Take care as some manufacturers have special insulated units that require the glass to go in a specific way. Such units carry a sticker that states something like *'this surface to face the outside'.*

Did you know?

Double-sided sticky rubber tape is a security device to prevent burglars removing the beads and taking the glass out.

Activity

In an area designated by your trainer, replace a double-glazed unit.

Knowledge refresher

1 Why is double glazing preferred to single glazing?

2 When replacing a double-glazed unit, why is it important that the glass is fitted in a specific way?

3 What are the three methods used when single glazing, and which is best?

4 How and why is double-sided sticky tape used when glazing?

What would you do?

You have been tasked with replacing a double-glazed unit. The original unit has condensation between the two panes of glass. What could have caused this? You remove the beading, but the unit still won't come out. What needs to be done? What should be done before fitting the new unit?

Repairing and replacing mouldings

Mouldings such as skirting or architrave rarely need repairing – usually only because of damp or through damage when moving furniture – and in most cases it is easier to replace them than repair them.

Architrave

To replace architrave, you simply remove the damaged piece and fit a new piece.

First check that the piece you are removing is not nailed to an existing piece. Then run a sharp utility knife down both sides of the architrave so that when it is removed it does not damage the surrounding decorations or remaining architrave.

Once the old piece is removed, clean the frame of old paint, give it a light sanding with sandpaper, then fit the new piece. Finally paint, stain or finish the new piece to match the rest.

Skirting board

Replacing skirting boards is slightly more difficult than replacing architrave. The way skirting boards are fitted could mean that the board you wish to replace is held in place by other skirting boards. Rather than remove the other skirting boards, you should cut or drill a series of holes in the middle of the board, splitting it in two so that you can remove it that way. Again, running a utility knife along the top of the skirting board will avoid unnecessary damage to surrounding decorations. Once the old skirting has been removed, the new piece can be fitted and finished to match the existing skirting.

You can replace other mouldings such as picture and dado rail in the same way as skirting boards, taking care not to damage the existing decorations.

Repairing a door that is binding

One of the most common problems that occur with doors is that they **bind**. A door can bind at several different points, as you can see in the illustration below.

These binding problems are simple to fix.

- If the door is binding at the hinges, it is usually caused by either a screw head sticking out too far or by a screw that has been put in squint. In these cases, screw the screw in fully or replace it straight. If the hinge is bent, fit a new one.

- If the door is binding on the hanging stile, it may not have been back-bevelled when hung. In this case, take the door off, remove the hinges, plane the door with a back bevel and then re-hang it. Alternatively, it may be expansion or swelling due to changes in temperature that is causing the problem. In this case, take off the door, plane it and re-hang it.

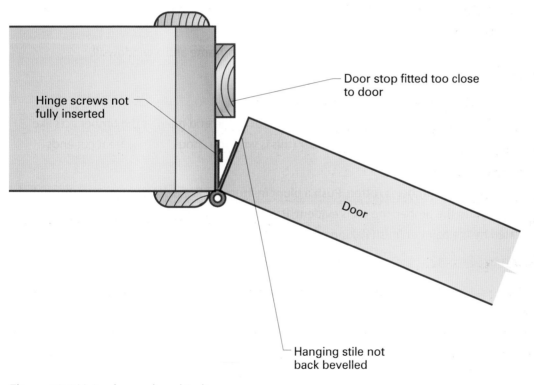

Hinge screws not fully inserted

Door stop fitted too close to door

Door

Hanging stile not back bevelled

Figure 10.6 Main places where binding occurs

- If the door is binding at the stop on the hanging side, the fitter may not have left a 1–2 mm gap between the stop and the door to allow for paint, in which case the stop will have to be moved. If there is a doorframe rather than a door lining, there is no stop to remove: simply plane the rebate using a rebate plane.

Activity

In an area designated by your trainer, repair a door that is binding.

Repairing damaged windowsills and doorframes

Provided they are painted or treated from time to time, exterior doorframes and windows should last a long time, but certain areas of frames and windows are more susceptible to damage than others. The base of a doorframe is where water can be absorbed into the frame; with windows, the sill is the most likely to suffer damage as water can sit on the sill and slowly penetrate it.

With damaged areas like these, the first thing to consider is whether to repair the damage or replace the component. With doorframes, replacement may be the best option but you must take into account the extra work required to make good: repairing the frame may cost less and require less work. With a rotten windowsill, replacement involves taking out the entire window so that the new sill can be fitted, so repair may be the better option.

The choice usually comes down to a balance of cost and longevity: as the experienced tradesperson, you must help the client choose the best course of action.

In this section we will look at repairing the base of a doorframe and a windowsill.

Repairing a doorframe

The base of the doorframe is most susceptible to rot as the end grain of the timber acts like a sponge, drawing water up into the timber (this is why you should always treat cut ends before fixing).

First check that the timber is rotten. Push a blunt instrument like a screwdriver into the timber: if it pushes into the timber, the rot is evident; if it does not, the frame is all right. Rot is also indicated by the paint or finish flaking off, or a musty smell.

If there is rot, repair it by using a splice, as follows.

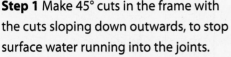

(a) make cuts in the frame

Decay or defect

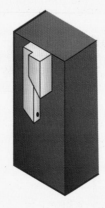

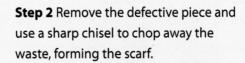

(b) remove the defect

Step 1 Make 45° cuts in the frame with the cuts sloping down outwards, to stop surface water running into the joints.

Step 2 Remove the defective piece and use a sharp chisel to chop away the waste, forming the scarf.

(c) Scarfed splice

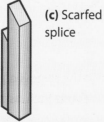

(d) Splice complete

Step 3 Use either a piece of similar stock or, if none is available, a square piece planed to the same size and shape as the original, making sure the ends are thoroughly treated before fitting to prevent a recurrence.

Step 4 Fix the splice in place, dress the joint using sharp planes and make good any plaster or render.

Figure 10.7 Stages in doorframe splicing

Activity

In an area designated by your trainer, repair a damaged doorframe.

This method can also be used to repair doorframes damaged by having large furniture moved through them, but the higher up the damage on the frame, the more likely the frame will be need to be replaced rather than repaired.

Repairing a windowsill

As with a doorframe, this method uses a splice, as follows:

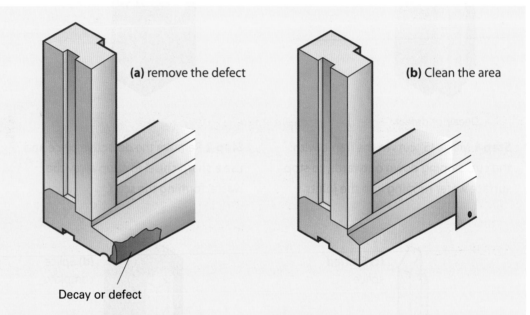

(a) remove the defect

Decay or defect

(b) Clean the area

Step 1 Make a cut at 45 degrees, then carefully saw along the front of the windowsill to remove the rotten area.

Step 2 Clean the removed area using a sharp chisel and then, as before, use either a piece of similar stock or, if none is available, a square piece planed to the same size and shape as the removed area.

Step 3 Give all cut areas a thorough treating with preservative, then attach the splice to match the existing windowsill shape and paint or finish to match the rest.

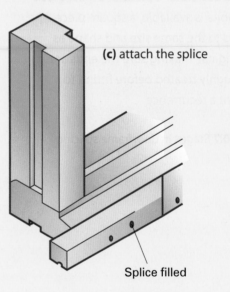

(c) attach the splice

Splice filled

Activity

In an area designated by your trainer, repair a damaged windowsill.

Figure 10.8 Stages in windowsill splicing

Knowledge refresher

1   Give a reason why mouldings may need to be replaced.

2   Define the term 'binding' in relation to doors.

3   How can you check if a doorframe is rotten?

4   What is the correct term for cutting in a repair piece at 45°?

What would you do?

You have been asked to give a quote for repairing the base of a doorframe. The client asks you what is best: repair or replace? What advice would you give? What factors do you need to take into consideration?

Replacing window sash cords

As you saw in Chapter 7, box sash windows use weights attached to cords that run over a pulley system, to hold the sashes open and closed. The sash cord will eventually break through wear and tear, and will need to be replaced. This is not a large job so, if one cord needs attention, it is most cost-effective to replace them all at the same time.

There are various ways to replace sash cords. The following method is quick and easy and is done from the inside.

Step 1 Remove the staff bead, taking care not to damage the rest of the window. Cut the cords supporting the bottom sash, lowering the weight gently to avoid damaging the case. Take out the bottom sash, removing the old nails and bits of cord, and put to one side.

Step 2 Pull the top sash down and cut the cords carefully. Remove the parting bead and then the top sash, again removing the old nails and bits of cord, and put to one side.

Step 3 Remove the pockets and take out the weights, removing the cord from them and laying them at the appropriate side of the window.

Step 4 Slide the top sash into place at the bottom of the frame and, using chalk, mark the position of the sash cord onto the face of the pulley stiles (see distance A in Figure 10.9). Mark the bottom sash in the same way (distance B). Put the sashes to one side.

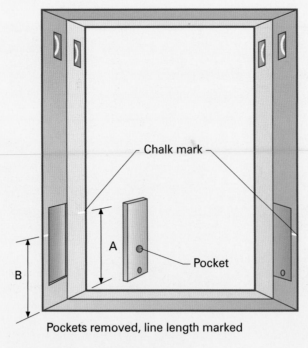

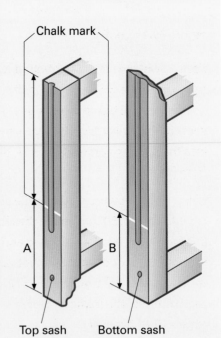

Figure 10.9 Marking sash lengths

Step 5 Next attach a **mouse** to the cord end, making sure it is long enough for the weighted end to reach the pocket before the sash cord reaches the pulley.

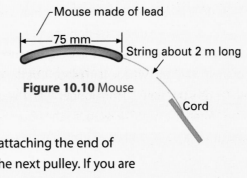

Figure 10.10 Mouse

Step 6 Now cord the window. If only one cord is to be replaced, you can do this by feeding one pulley, then cutting the sash to length, re-attaching the end of the mouse to the sash cord and then feeding the next pulley. If you are replacing more than one cord, it is more efficient to feed the pulleys in succession and cut the cords after. In Figure 10.11 you can see one method for cording the window.

Definition

Mouse – a piece of strong, thin wire or rope with a weight such as a piece of lead attached to one end, used to feed the sash cord over the pulleys in a box sash window

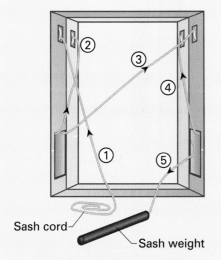

Figure 10.11 Cording a window

Feed the cord through the top left nearside pulley (1), out through the left-hand pocket, in through the top left far side pulley (2), then back through the same pocket. Then feed it through the top right nearside pulley (3), out through the right-hand pocket, in through the top right far side pulley (4) and back out of the right-hand pocket. Remove the mouse and attach the right-hand rear sash weight to the cord end (5).

Step 7 Working on the right-hand rear pulley, pull one weight through the pocket into the box and up until it is just short of the pulley. Lightly force a wedge into the pulley to prevent the weight from falling.

Pull the cord down the pulley stile and cut it to length, 50 mm above the chalk line (when the window is closed and the weight falls back down to the bottom of the box, the weight will stop 50 mm from the bottom of the frame as shown in Figure 10.14 overleaf).

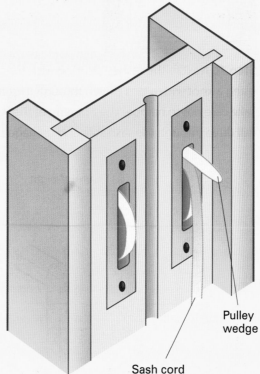

Figure 10.12 Wedged pulley

Step 8 Tie the other right-hand weight to the loose end of the cord. Still on the right-hand near-side, pull the cord again so that the weight is just short of the pulley and wedge it in place. Cut the cord to length, this time 30 mm longer than the chalk line (when the window is closed the bottom sash weight will be 30 mm away from the pulley, as shown in Figure 10.14).

Step 9 Fix the right-hand pocket back in place, then repeat Steps 7 and 8 for the left-hand side. Now all the pulleys are wedged and all the cords cut to length.

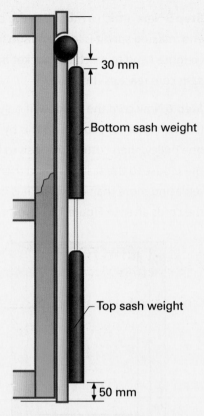

Figure 10.14 Sash weights at 50 mm and 30 mm

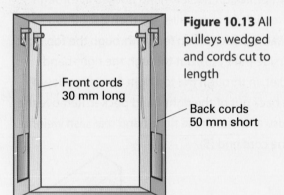

Figure 10.13 All pulleys wedged and cords cut to length

Front cords 30 mm long

Back cords 50 mm short

Step 10 Next fit the sashes, starting with the top sash. Fix the cord to the sash by either using a knot with a tack driven through it, or a series of tacks driven through the cord. Take care not to hamper the opening of the window, and do not use long nails as they will drive through the sash stile and damage the glass.

Activity

In an area designated by your trainer, replace damaged sash cords.

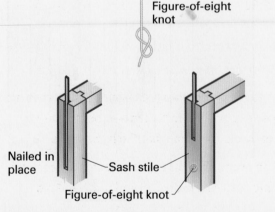

Figure-of-eight knot

Nailed in place

Sash stile

Figure-of-eight knot

Figure 10.15 Two methods for fixing cord to sashes

Step 11 Once the cord is fitted to both sides of the sash, slide the sash into place and carefully remove the wedges. Now test the top sash for movement and then re-fit the parting bead to keep the top sash in place. Now fit the bottom sash as in Step 10, test it and re-fit the staff beads, to secure the bottom sash in place.

Now test the whole window by sliding both sashes up and down. Finally, touch up any minor damage to the staff and parting beads.

Knowledge refresher

1 Name the component attached to the end of a sash cord, used to help feed the cord through the pulleys.

2 Describe two methods of fixing sash cord to sashes.

3 State an alternative to sash cords and weights.

What would you do?

You have been asked to repair a box sash window that is not opening properly. What might the cause be? What can be done to rectify it? If there are major problems, what is it best to do: repair or replace?

Repairing structural timbers

The maintenance of structural timbers is vital, as they will almost certainly be carrying a load: joists carry the floors above, while rafters carry the weight of the roof. Because of this, structural repairs should be carried out by qualified specialists, so this section will give a brief understanding of the work that is involved.

Joists

Joist ends are susceptible to rot: they are close to the exterior walls and can be affected in areas like bathrooms if the floor gets soaked and does not dry out. Joists can also be attacked by wood-boring insects. For both rot and insect damage, the repair method is the same.

Shore the area, with the weight spread over the props, then lift the floorboards to see what the problem is (for the purposes of this example, we will use dry rot) and throw the old floorboards away.

Before making any repairs, you need to find and fix the cause of the problem – otherwise the same problems will keep on arising. Rot at the ends of joists is usually caused by poor ventilation: the cavity or the airbricks may be blocked, for example.

Next cut away the joists allowing 600 mm into sound timber, and treat the cut ends with a suitable preservative.

New pre-treated timbers should be laid and bolted onto the existing joists with at least 1m overlap. New timbers should ideally be placed either side of the existing joist, and in place of the removed timber.

Any new untreated timbers should be treated with a suitable preservative, before or after laying. Now slowly remove the props and make good as necessary.

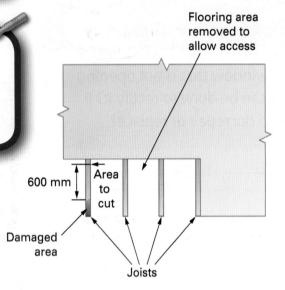

Figure 10.16 Cut away the rotten area of the joist

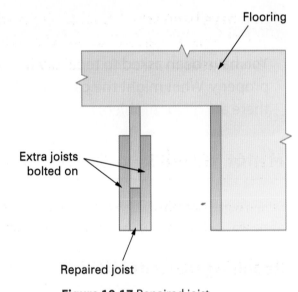

Figure 10.17 Repaired joist

Rafters

Rafters are susceptible to the same problems as joists, and especially to rot caused by a lack of ventilation in the roof space. Again the problem needs to be remedied before any repairs start.

Rafters are replaced or repaired in the same way as joists. They are easier to access, but shoring rafters can be difficult. In some cases it is best to strip the tiles and felt from the affected portion of the roof before starting.

For trussed roofs you should contact the manufacturer: trusses are stress graded to carry a certain weight, so attempting to modify them without expert advice could be disastrous.

Knowledge refresher

1 Why should trussed rafters only be repaired by a specialist?

2 Why should there be at least 1 m overlap when bolting extra joists onto a repaired joist?

What would you do?

You have been asked to look at joists that appear to be damaged by rot. The client wants to know what has caused it. What has caused it? What needs to be done to rectify it?

Remember

Attempt only minor repairs unless you are fully trained. Otherwise you could end up doing more harm than good!

Minor repairs to plaster and brickwork

During repair work there is always a chance that the interior plaster or brickwork may get damaged. Rather than call in a specialist most tradespeople will repair the damage themselves.

Repairing plasterwork

Plaster damage most often occurs when windows or doorframes are removed – the plaster cracks or comes loose – and is simple to fix with either ready-mixed plaster (ideal for such tasks) or traditional bagged plaster, which is cheaper.

Bagged plaster needs to be mixed with water and stirred until it is the right consistency. Some people prefer a thinner mix as it can be worked for longer, while others prefer a consistency more like thick custard as it can be easier to use.

Whichever type of plaster you use, the method of application is the same.

First brush the affected area with a PVA mix, watered down so it can be applied by brush. This acts as a bonding agent to help the plaster adhere to the wall.

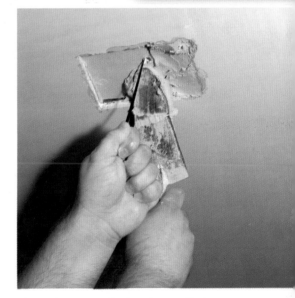

Plaster being applied and trowelled

Next use a trowel or float to force the plaster into the damaged area. If it is quite deep you may have to part-fill the area and leave it to set, then put a finish skim over it. Filling a deep cavity in one go may result in the plaster running, leaving a bad finish.

When the plaster is almost dry, dampen the surface with water and use a wet trowel to skim over the plastered area, leaving a smooth finish. Once the plaster is dry, the area can be redecorated.

Repairs to exterior render are done in the same way, using cement instead of plaster.

Repairing brickwork

With repairs to brickwork, one of the first problems is finding bricks that match the existing ones, especially in older buildings. After that, the method is as follows.

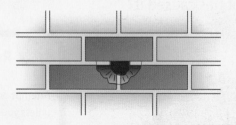

(a) Step 1

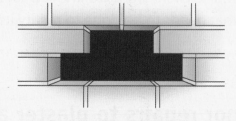

(b) Step 2

Step 1 Make a hole in the centre of the area, to allow the removal of the bricks. Bricks rarely come out whole, but care must be taken not to damage the surrounding bricks.

Step 2 Remove the bricks and carefully clean away the old mortar using a cold chisel.

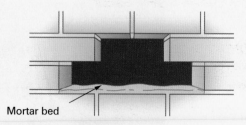

Mortar bed

(c) Step 3

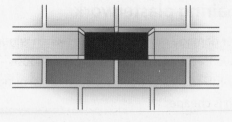

(d) Step 4

Step 3 Lay a mortar bed on the base of the opening.

Step 4 Place the first two bricks, making sure mortar is applied to all the joints.

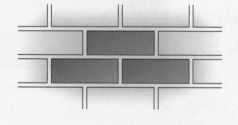

(e) Step 5

Step 5 Place the last brick, again making sure there is mortar between all the joints, then use a pointing trowel to point in the new bricks to match the others.

Figure 10.18 Stages in repairing brickwork

FAQ

Do I have to use putty or silicone to re-fix a pane of glass or can I use an adhesive?

Like other substances, glass expands and contracts slightly depending on the weather. This means that silicone or putty must be used as they remain flexible even when set. Adhesive sets rigid so, if any expansion or contraction occurs, the glass will break.

Can I put a double-glazed unit in a window or door that originally contained a single pane of glass?

Yes, as long as the rebate is wide enough to accept the unit and beading.

Can I do a scarf repair on skirting or architrave?

Yes, but it is often easier to replace mouldings than to repair them.

Should I go on a plastering course to help with repairs?

This is not necessary as the repairs you would be doing are not large, but if you feel that it would benefit you, yes.

Knowledge refresher

1 What is the first step to take before applying plaster?

2 What should be done when plastering a deep cavity?

What would you do?

You have been asked to make plaster repairs around a new window that has been fitted. The plaster keeps slipping and you can't get an even finish. What can be causing this? What can be done to rectify it?

chapter 11

Woodworking machines

OVERVIEW

Woodworking machines are vital in producing components accurately and safely. The machining of timber has become more prevalent in recent years. Technology has allowed the building of machines that can transform stock timber into any size or shape, making the woodworker's job far easier. Every carpenter or joiner will at some stage come across a woodworking machine, whether it be a combination planer in a workshop or a table saw on site. It is important that you understand the principal uses, safety considerations and set-up of woodworking machines.

This chapter will cover:

- general safety of woodworking machines
- table saws
- planers (surface, thickness and combination)
- band saws
- mortise machines.

General safety of woodworking machines

Compared to other industries, woodworking accounts for a large proportion of accidents. Woodworking machines often have high-speed cutters, and many cannot be fully enclosed owing to the nature of the work they do.

The use of woodworking machines was originally governed by the Woodworking Machine Regulations 1974. The introduction of the Provision and Use of Work Equipment Regulations (PUWER) 1992 superseded the 1974 Regulations, although regulations 13, 20 and 39 were still in use until the PUWER regulations were updated in 1998.

The PUWER regulations are explained in more detail in Chapter 1, but below are a few of the items relating to the safe use of all woodworking machines. Safety regulations relating specifically to a particular machine are noted in dedicated sections.

Safety appliances

Safety appliances such as push sticks/blocks and jigs must be designed so as to keep the operator's hands safe. More modern machines use power feed systems, eliminating the need for an operator to go near the cutting action. Power feed systems should be used wherever possible; in the absence of a power feed system, the appropriate push sticks/blocks must be used.

Working area

An unobstructed area is vital for the safe use of woodworking machines. The positioning of any machine must be carefully thought through to allow the machine to be used as intended. In a workshop environment, where there are several machines, the layout should be arranged so that the materials follow a logical path. Adequate access routes between machines must be kept clear, and there should also be a suitable storage area next to each machine to store materials safely without impeding the operator or others.

Floors

The floors around machines must be kept flat and in a good condition, and must be kept free from debris such as chippings, waste wood and sawdust. Any electricity supply, dust collection ductwork, etc. must be run above head height or set into the floor in such a way that does not create a trip hazard. Polished surfaces must be avoided and any spills must be mopped up immediately. Non-slip matting around a machine is preferred, but the edges must not present a trip hazard.

Did you know?

The PUWER regulations cover all working equipment, not just woodworking machines.

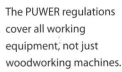

Activity

Check your workshop area to see if it meets current regulations.

Lighting

All areas must be adequately lit, whether by natural or artificial lighting, to ensure that all machine set-up gauges and dials are visible. Lighting must be strong enough to ensure a good view of the machine and its operations, and lights must be positioned to avoid glare and without shining into the operator's eyes.

Heating

The temperature in a workshop should be neither too warm nor too cold, and the area should be heated if needed. A temperature of 16 degrees C is suitable for a workshop.

Controls

All machines should be fitted with a means of isolation from the electrical supply separate from the on/off buttons. The isolator should be positioned so that an operator can access it easily in an emergency. Ideally there should be a second cut-off switch accessible by others in case the operator is unable to reach the isolator. Machines must be fitted with an efficient starting/stopping mechanism, in easy reach of the operator. Machines must always be switched off when not in use and never left unattended until the cutter has come to a complete standstill.

Braking

All new machines must be fitted with an automatic braking system that ensures that the cutting tool stops within 10 seconds of the machine being switched off. Older machines are not required to have this, but PUWER states that all machines must be provided with controls that bring the machine to a controlled stop in a safe manner. The approved code of practice calls for employers to carry out a risk assessment to determine whether any machine requires a braking system to be fitted, and includes a list of machines for which braking will almost certainly be needed. If you are unsure, contact your local Health and Safety Executive.

Dust collection

Woodworking machines must be fitted with an efficient means of collecting the dust or chippings produced during the machining process.

Training

No person should use any woodworking machinery unless they have been suitably trained and deemed as competent in the use of that machine.

Maintenance

The machines should be maintained as per the manufacturer's instructions and should be checked over prior to use every day, with the inspection and any findings recorded in a log.

Knowledge refresher

1   What regulations govern the use of woodworking machines?

2   What regulations did the current ones replace?

3   What is a suitable temperature for a workshop?

4   Under the new regulations, how long should it take a machine to come to a complete stop once the stop button has been pushed?

What would you do?

You are using the circular saw when a strange noise starts up. You switch the machine off and isolate it. Your boss tells you to fix it yourself. What should you do? What could the consequences be?

Table saws

Table saws – also known as rip saws, circular table saws and bench saws – come in a range of shapes and sizes.

Flatting (note that the guards have been adjusted for photographic purposes)

The main functions of a table saw are flatting (cutting the timber to the required width) and deeping (cutting the timber to the required thickness).

With advances in technology and a new variety of saw blades available, the table saw is capable of other tasks including creating housings. Here we will deal with the more traditional table saw, covering:

- table saw parts
- saw blades
- machine set-up
- safe operation.

Table saw parts

Activity

Set up a circular saw ready for use.

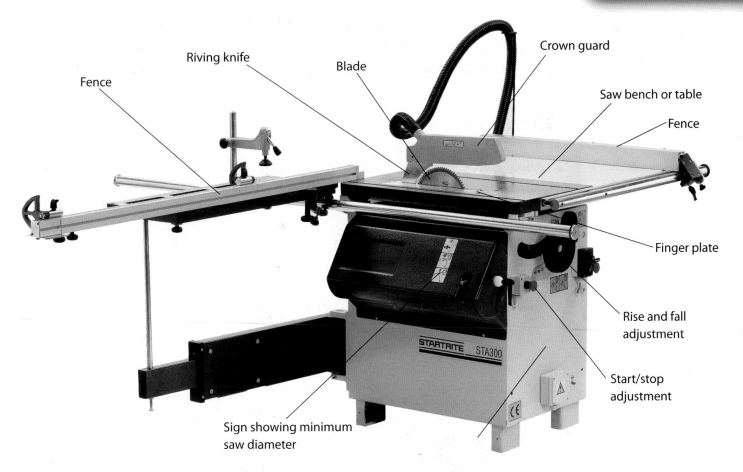

Parts of a traditional table saw

- **Saw bench or table** – the table or bench on which the saw is fitted. This should be extendable via rollers or extra out-feed tables to allow larger materials such as sheet timber to be machined

- **Blade** – the cutting tool

- **Crown guard** – a guard suspended over the top of the blade. It must be adjustable to ensure that as much of the blade is guarded as possible

- **Riving knife** – thicker than the blade, this acts as a spreader to prevent the cut from closing, which could cause binding

- **Fence** – a guide to give straight and accurate cutting, which should be accurately labelled and easily adjustable

- **Fingerplate** – a cover piece that sits over the spindle of the saw

- **Rise and fall adjustment** – a wheel used to adjust the height of the blade

- **Start/stop button** – this should be clearly labelled and within easy reach of the operator

- **Sign showing minimum saw blade diameter** – a legal requirement.

Activity

Sketch a circular saw blade, showing the pitch, clearance angle, gullet, hook, etc.

Activity

Change the blade on a circular saw.

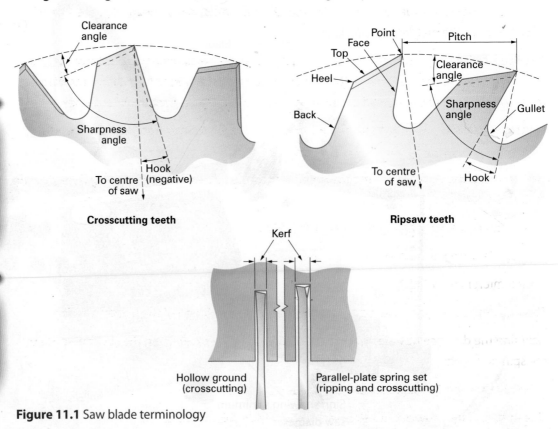

Figure 11.1 Saw blade terminology

Saw blades

Several different types of saw blade are available, but here we will look at the standard types.

- **Pitch** – the distance between two teeth

- **Hook** – the angle at the front of the tooth. A positive hook is required for ripping, with an angle of 20–25 degrees for softwoods and 10–15 degrees for hardwood
- **Clearance angle** – ensures that the heel clears the timber when cutting
- **Top** – the angle across the top of the tooth
- **Set** – the amount each tooth is sprung out to give clearance and prevent the saw from binding
- **Gullet** – the space between two teeth, which carries away the sawdust
- **Kerf** – the width of the saw cut (not the width of the blade).

Tungsten carbide-tipped (**TCT**) blades are now preferred as they stay sharper for longer and do not have a set.

Calculating the rim or peripheral speed of a blade

The relation of the saw blade size to the speed of the spindle is crucial for the safe running of the saw (this is why the minimum size of the blade fitted must be shown clearly on the machine). A blade with a rim/peripheral speed of 100 metres per second can become stressed, causing a dangerous situation. A slow rim speed, on the other hand, means that the saw will struggle to cut the material.

To calculate the rim/peripheral speed, we must first know the diameter of the saw blade and the spindle speed. The diameter of the saw blade can be measured or found from the blade manufacturer; the spindle speed can be found from the saw manufacturer's handbook.

For this example we will use a saw blade with a diameter of 600 mm and a spindle speed of 1600 revolutions per minute.

First work out the distance around the rim (the circumference of the blade), using the following calculation:

Circumference = π × diameter (where π is 3.142)

Circumference = 3.142 × 0.600

Circumference = 1.8852

This equals the distance travelled by 1 tooth in a single revolution of the blade.

Next find the distance travelled every minute by 1 tooth, by multiplying the circumference by the spindle speed:

1.8852 × 1600 = 3016.32 metres per minute

Finally, to find the answer in metres per second, divide this answer by 60:

3016.32/60 = 50.272 metres per second.

A rim/peripheral speed of 50 metres per second is considered suitable.

If you are unsure about which blade size to fit, contact the manufacturer.

Machine set-up

The set-up of any table saw depends on the manufacturer and the type and model of saw, so be sure to check the manufacturer's handbook prior to using any machine. Whatever the make or model of your saw, there are certain safety measures you must take before use. You must ensure that:

- the machine is isolated from the power source prior to setting up
- all blades are in good condition, securely fixed and running freely
- all guards, guides, **jigs** and fences are set up correctly and securely
- suitable push sticks are available and at hand ready to use
- suitable PPE is available.

Safe operation

The operation of each table saw varies slightly depending on the task you are doing, but the basic operating principles remain the same.

First ensure that the checks have been done and the machine is set up safely. Wear the appropriate PPE and have a push stick at hand ready for use.

The machine can then be started and allowed to get up to speed before you offer the timber to the blade. Stand at the side of the piece being sawn and not directly behind it, as **kickback** may occur and can cause injury.

Using the fence/jig as a guide, slowly ease the timber into the blade using a steady amount of pressure: applying too much pressure will cause the blade to overheat, leaving burn marks on the face of the timber and even causing the blade to wobble, resulting in kickback. The right amount of pressure varies from material to material and between sizes of material but, generally, if you hear the saw straining or if there is a burning smell from scorching timber, you are applying too much pressure.

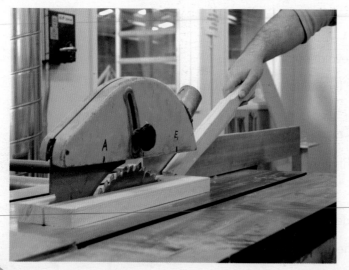

Keep the timber hard against the fence to ensure accuracy, and continue to feed it into the saw until there is no less than 300 mm remaining to be cut. The last 300 mm or more must be fed through using a push stick. The push stick should also always be used to remove the cut piece between the saw blade and the fence and any other off-cuts.

Correct use of a push stick (N.B. guards have been adjusted for this photograph to reveal the blade).

FAQ

What is binding?

Binding is where the timber being sawn closes or 'binds' over the blade so that contact with the face of the blade is made. This causes the timber to be thrown back at the operator ('a kickback').

What causes binding?

Binding can be caused by many different things including overexerting the blade by applying too much pressure and using a blunt or damaged blade. However, the most common cause of binding is high moisture content in the timber. Timber with a high moisture content has a tendency to move while being machined, which in some cases causes the kerf to close at the rear of the riving knife, applying pressure to the blade and causing binding.

How can I prevent the timber from binding when machining timber with high moisture content?

The best way is to have a second person who stands at the rear of the saw blade (well away from the blade) and drives timber wedges into the kerf to keep it from closing.

What is the best way to rip long lengths of timber?

Either have a long out-feed table or an extra worker at the rear to help guide the timber through the saw and support the weight of the timber as it passes through.

The timber that I am ripping down comes out a different measurement from the one I set the saw to. Why is this?

The gauge on the fence is out and must be recalibrated.

FAQ

The saw is not cutting straight. Why is this?

There are two main causes. Either the fence is not secured properly and is moving when you are feeding the timber through, or the timber is not being held tight against the fence throughout the operation.

What length does a push stick need to be and why?

A push stick must be between 300 and 450 mm. Any less than 300 mm and your hand will be too close to the operating blade; any more than 450 mm and you are too far away to have proper control. Also, the longer the push stick the more chance there is of it springing out of place.

Knowledge refresher

1   What are the two main functions of a saw table?

2   What is the purpose of a crown guard?

3   What is the pitch of a saw blade?

4   What is the fingerplate on a circular saw?

5   What is the importance of peripheral speed?

What would you do?

You have been asked to change the blade on a circular saw. You work out the peripheral speed, and it comes out at almost 100 mps. What are the hazards of using this? What causes these hazards? What should you do?

Planers

A hand-fed power planer is another essential tool in a woodwork shop as it can do in seconds what a craftsman would take hours to do by hand. This sort of planer is in essence the same as a portable power planer, but on a much larger scale and more accurate. All planers comprise an in-feed table, an out-feed table and a cutter block, into which either two or three cutting blades are housed.

Three main types of planer are available:

- surface planer
- thickness planer
- combination planer.

Activity

Use planers to face, edge and finish materials to a desired thickness.

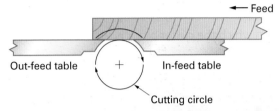

Figure 11.2 Cutter block, with out-feed and in-feed tables

The combination planer is by far the best machine to have, as it can carry out all the operations of both surface and thickness planers. Given that the combination planer is essentially a combination of the other two, we will only look at the surface and thickness planers here.

Surface planer

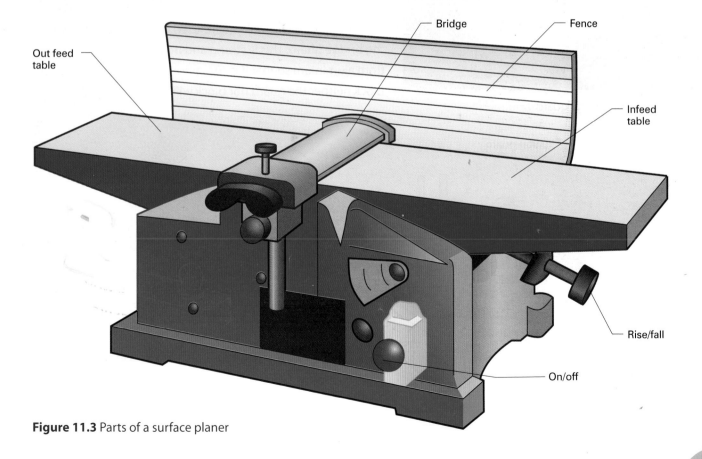

Figure 11.3 Parts of a surface planer

As its name suggests, the surface planer creates a smooth and even surface on the piece of timber. It is most commonly used for two main operations:

1. **facing** – when the timber is planed flat and even on the widest side of the timber

2. **edging** – when the timber is placed flat and even on the narrowest side of the timber.

With the use of a fence the surface planer can also be set up to create rebates and splayed or angled pieces.

The surface planer works by passing the timber smoothly over an in-feed and out-feed table, in between which the cutter block is situated. The height of the in-feed table is adjustable to regulate the amount of timber removed in a single pass.

The in-feed and out-feed tables are machined to be perfectly flat. When facing, provided the timber is held firmly to the surface of the out-feed table, a perfectly flat surface is produced. Edging is done by placing the faced side of the timber against the fence and running it over the machine.

The surface planer is fed mostly by hand, so great care must be taken to avoid your hands coming into contact with the cutter block.

The main way of protecting the user is through the use of suitable guards. The main guard used on a surface planer is the bridge guard. This is strong and rigid, and is usually made from aluminium so that if it comes into contact with the cutter block neither the guard nor block will disintegrate. Telescopic

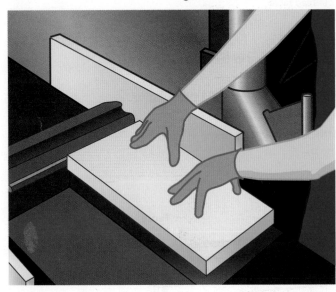

Figure 11.4 Telescopic bridge guard

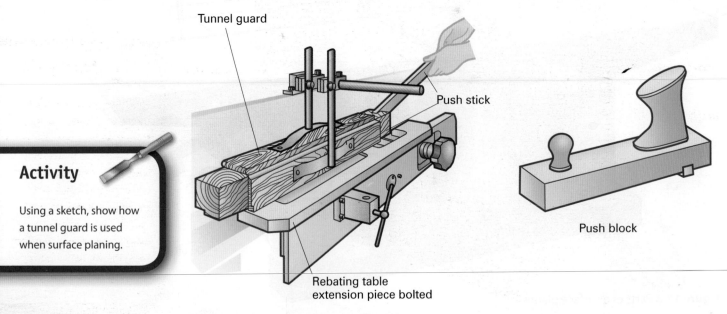

Tunnel guard

Push stick

Push block

Rebating table extension piece bolted

Figure 11.5 Tunnel guard and push block

bridge guards are advisable as they can cover the full length of the cutter block; they must be wider than the cutter block.

Tunnel guards can also be used in conjunction with a push stick; push blocks should also be used when appropriate.

There can be a tendency to feed timber over the cutter block too fast, which will leave a bad finish; to produce a good finish, it is best to go slowly.

Thickness planer

The thickness planer planes the timber to the required width or thickness, usually after the surface planer has faced and edged the timber. The thickness planer can also be used with a variety of jigs to create simple mouldings such as window beads.

The thickness planer has a power feed system, usually in the form of four rollers. Two idle rollers are fitted at the bottom on the rise and fall table, and the two rollers above these are

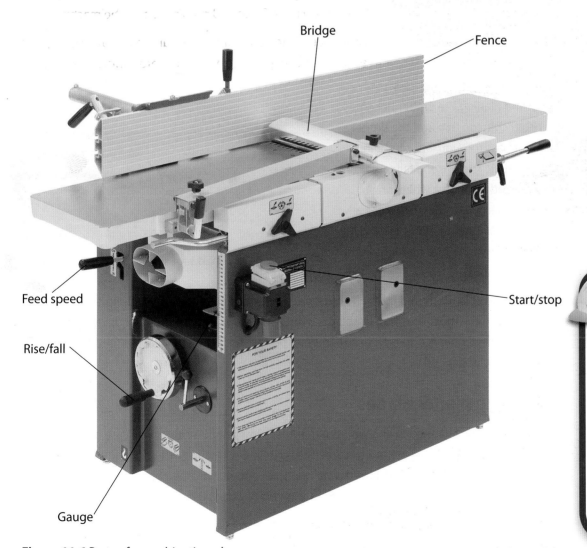

Figure 11.6 Parts of a combination planer

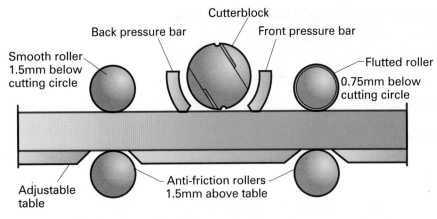

Figure 11.7 Section through a thickness planer

driven rollers, which propel the timber through the machine. The first upper roller is usually serrated so that it can grip the sawn timber better, while the second power roller is smooth so as to not damage the finished surface. To prevent kickback both upper rollers are spring-loaded, as are the two pressure bars. On most modern machines the serrated roller and front pressure bar are made in sections to allow more than one piece to be fed into the machine at the same time.

Operating the thickness planer is very simple: the timber is fed into the machine at one end and comes out planed to the required thickness at the other. You must be careful to ensure that the timber being fed in is of an appropriate thickness: too thick and planing the timber becomes very difficult; too thin and the timber will just pass through without being planed at all. The rise and fall table can be adjusted through a wheel or, on more modern machines, electronically via a button. A depth gauge should be in easy view so that the required depth can be set easily.

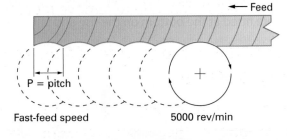

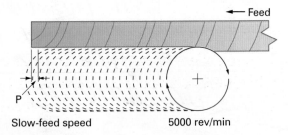

Figure 11.8 Pitch marks for slow- and fast-speed feeds

Activity

Set up a thickness/combination planer ready for dressing materials to correct thickness.

The thickness planer should have an adjustable roller-feed speed setting to help create a better finish. One of the most common problems with planing machines is pitch marks, where the surface of the timber is rippled. Pitch marks can be caused by having a fast in-feed speed or a slow cutter block speed.

Changing planer blades

Activity

Change the blades on a planing machine.

Blades should be changed in line with the manufacturer's guidance. Special care must be taken to ensure that new blades are fitted correctly and in line with each other. Wrongly aligned blades may result in a poor finish: with one blade set above the other, only one blade is doing all the cutting, resulting in pitch marks. Poorly set blades may also pose a safety risk as they may shatter or chip.

FAQ

There is a small chip out of the blade on the surface planer I am using. Do I need to change the blades?

No. The fence adjustment should be moved so that you are planing in the area without the chip in it. Replacing blades every time there is a small chip can be an expensive practice.

How much material can I remove on a single pass with a surface planer?

It is not recommended to remove more than 3 mm in a single pass.

I find it difficult to get a good finish when surface planing long lengths of timber.

When machining long lengths, it is best to have two people: one at the front and the other at the back, supporting the weight.

What PPE do I need to wear when operating planing machines?

You should wear boots, gloves, goggles, ear defenders and, depending on the timber being machined, a dust mask. You must also ensure that you have no loose clothing that could get caught in the cutter block.

The thickness planer I am using is creating pitch marks. How can I solve this?

Check the machine's feed speed and adjust it so that the machine feeds more slowly.

How do I know if I can feed more than one piece of timber into the thickness planer at the same time?

There should be a label on the machine stating that only one piece can be machined at any time. If there is no label, check the manufacturer's instructions. If you are in any doubt, feed just one piece at a time.

Knowledge refresher

1   Name the three main types of planer.

2   Name three operations that can be carried out on a surface planer.

3   When planing, what can cause pitch marks?

4   What type of guards can be fitted to a surface planer?

What would you do?

You have been tasked with running 200 pieces of timber through a thickness planer to machine them to a desired finish size. The machine is a single-roller feed, meaning that you should only feed one piece of timber in at a time. The materials need 7 mm removed, which means that each piece should be fed into the machine three times. Your boss needs the job done quickly, and asks you to feed four pieces in at the same time, taking the full 7 mm off. What hazards can this create? What effect will this have on the finish? What should you do?

Band saws

The band saw is mainly used for curved or shaped work, but with the addition of fences a band saw can be used for ripping and crosscutting too. A band saw consists of an endless blade that runs around two wheels, with one wheel mounted above the other. The wheels are encased in the machine to protect the user, and an adjustable table sits between the wheels, which is where the cutting action occurs.

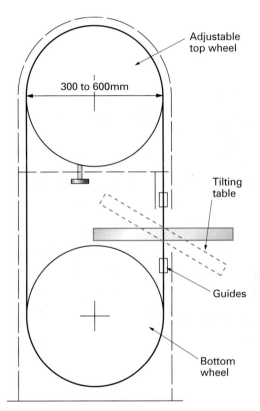

Figure 11.9 Wheels, table and blade for a band saw

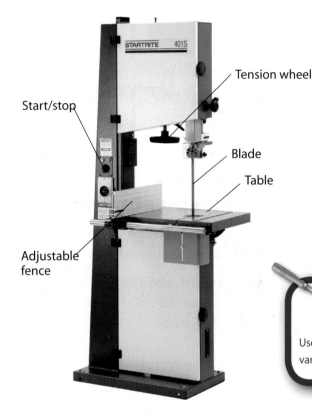

Figure 11.10 Parts of a band saw

Activity

Use a band saw for cutting a variety of materials.

Activity

Using a sketch, show how the guides at the top and bottom of a band saw are set up.

The bottom wheel of the band saw is driven directly by the motor, while the top wheel is driven by the blade. The top wheel is also adjustable to allow the blade tension to be set. Rubber tyres attached to the rims of the wheels help to stop the blade slipping, as well as preventing damage to the saw blade.

As well as running on the wheels, the blade is supported by saw guides situated above and below the saw table. The top guide is fitted with a guard to protect the user. The guides are fitted with thrust wheels, which prevent the saw blade being pushed back when the work is pressed against it. Blades come in a variety of sizes from 3 mm upwards: the smallest blades are for more intricate, curved work; the larger blades are for ripping large, sectioned timber.

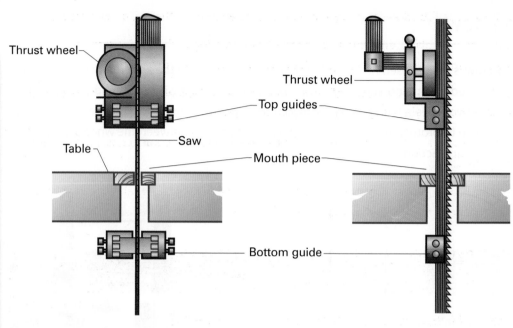

Figure 11.11 Guides

Setting up the band saw

As with all machines, the set-up depends on the manufacturer. Not every machine can do all tasks, so it is always best to refer to the manufacturer's handbook when setting up the machine.

Before use, it is important to check the tracking. If the wheels are not set correctly, the blade could come off the wheels or tilt backwards, running against the guide and damaging or even breaking the blade.

Using the machine

The machine's operations will be listed in the handbook. Always ensure the machine is set up correctly for the operation you need prior to use. When working intricate curves into timber with a band saw your hands can get close to the blade, so it is vital to use push sticks and the correct fence attachments. If a band saw blade does snap, there can be a loud bang (and it can be quite scary), but all band saws must be fitted with a brake designed to stop the machine as quickly as possible, preventing the wheels from continuing to drive the broken blade.

Changing the blade

If a blade does break, or the machine needs to be set up for a different use, the blade will need to be replaced. Again, the manufacturer's handbook should be followed, but here is a basic guide to changing a blade:

Step 1 Ensure that the machine is switched off and isolated from the power supply before making any adjustments.

Step 2 Open the top and bottom guard doors and move aside any other obstructions such as the blade guards and table mouthpiece.

Step 3 Lower the top wheel to remove the tension, then remove the old/broken blade.

Step 4 Fit the new blade, ensuring that the teeth are pointing in the right direction (downward in most cases), then readjust the tension so that it is set correctly.

Step 5 Set the tracking so that the blade is running true and will not come off the wheels. Spin the wheels a few times to check that the tracking is correct.

Step 6 Reattach the guards, etc., reset the thrust wheels and guides and close the top and bottom guard doors.

Step 7 Switch the machine on and let it run for a little while to ensure that it is tracked and tensioned properly before attempting to cut any materials.

Activity

Change the blade on a band saw.

FAQ

How do I know what tension to set the blade to?

There should be a sign or panel on the machine that tells you what tension each type of blade should be set to. If not, contact the manufacturer.

There is a smell of burning rubber when the machine is in use. What is causing this?

This is happening because the tracking is out, making the blade cut into the tyres.

There is a thin blade in the machine and I want to rip large timber. Do I have to change the blade?

Yes, the thin blade will not cope with the task and will break.

Do I have to throw out my broken or blunt blades?

No, they can be repaired. However, band saw blades are not expensive so it is often best to buy new.

There are sparks appearing when I use the saw. What causes this?

Sparks can occur when the thrust wheel is not set correctly or when the tracking or tension is not correct. Switch off the machine, isolate it and check the thrust wheel, tracking and tension.

Knowledge refresher

1 What are band saws mainly used for?

2 How should the tension of a band saw blade be checked?

3 What is the purpose of the tracking on a band saw?

What would you do?

You have been tasked with machining circular pieces on a band saw. You have changed the blade to a smaller blade to allow for the cutting of curves. You start the machine up and there is a smell of burning rubber coming from the machine. What has caused this? What can be done to rectify it? What would happen if you did not rectify it?

Mortise machines

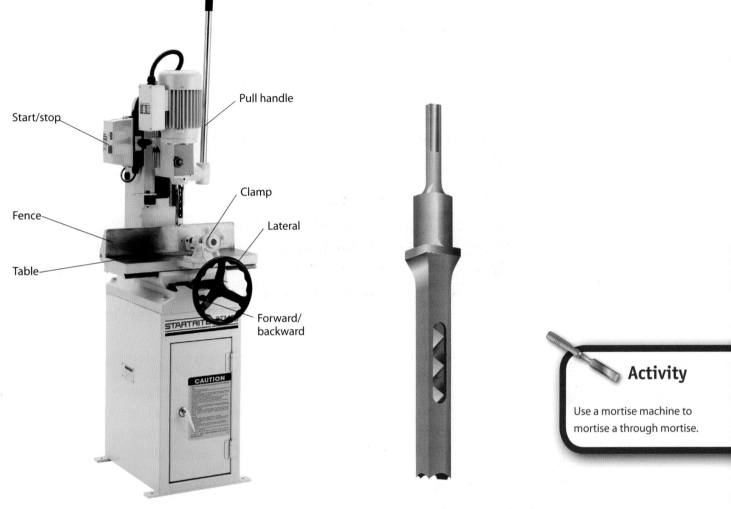

Figure 11.12 Parts of a mortise machine   **Figure 11.13** Hollow square chisel

Activity

Use a mortise machine to mortise a through mortise.

The hollow-chisel mortise machine is designed predominantly to chisel out mortises for mortise and tenon joints, but it can also be used to chisel out mortises for any purpose.

A mortise machine consists of a revolving auger housed inside a hollow square chisel (this comes in various sizes, producing a range of mortises). The auger cuts most of the timber, while the chisel squares the hole up. The machine is operated much like a pillar drill, with a lever forcing the cutting action into the work piece. When setting the machine up it is important to ensure that there is a clearance of 2–3 mm between the tip of the auger and the chisel: this prevents the tips of the auger and chisel overheating.

The work piece sits on a table and is clamped in place to prevent it from moving. The table can be moved forwards and backwards as well as from side to side; this means more than one hole can be drilled without removing the timber from the clamp.

Setting up and using the mortise machine

The mortise machine should be set up, and any tooling changed, as per the manufacturer's handbook. If the tooling has been changed, it is a good idea to run a test piece to ensure that the mortise is set up square; this prevents stepped mortising.

Once the machine has been set up correctly, the work piece can be clamped in place and any lateral or sideways adjustments can be made so that the chisel cuts in the correct place. When mortising full mortises, it is good practice to go only a little over halfway down the depth, then turn the timber over and complete the mortise from the other side; this avoids splitting out the bottom of the timber.

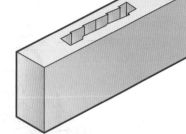

Figure 11.14 Stepping mortise caused by chisel not being set correctly

Safety tip

When machining timber of long lengths, make sure that they are adequately supported on any overhanging ends to prevent the work piece from lifting.

Activity

Set up a mortise machine.

Activity

Change the tooling on a mortise machine.

FAQ

How do I know if I have mortised halfway through?

Most machines come with a depth stop that can be set to any depth.

I want to mortise a 21 mm mortise but the only chisels available are 18 mm or 22 mm. What do I do?

You cannot use the 22 mm chisel, as it is too big. Instead, use the 18 mm chisel and adjust the table laterally to complete the mortise.

Knowledge refresher

1   What is the purpose of a chisel on a mortise machine?

2   What causes stepped mortises?

What would you do?

You mortise the first piece of a series of through mortises for a frame and, when you remove it from the machine, the back of the mortise is split. What could have caused this? What could be done to prevent it?

chapter 12

Mark, set out and manufacture joinery products

OVERVIEW

While the carpenter is generally employed on site carrying out first and second fixing and carcassing, the joiner is most likely to be found in a workshop manufacturing the components that the carpenter will fit on site. Whichever qualification you follow, you will need a good understanding of how basic components such as stairs are made.

This chapter will first cover general points about:

- setting out and marking.

Next, you will find details of the setting out, marking and manufacture of:

- windows
- doors
- stairs
- basic units.

Setting out and marking

The joints used in the manufacture of joinery components are covered fully in *Carpentry and Joinery NVQ and Technical Certificate Level 2*. This section will give a brief recap of the marking and setting out process.

Whatever you are setting out, it is best to start with a plan or a drawing. For joinery products, this is usually done to full scale, on a thin sheet of board (plywood, hardboard, etc.) better known as a setting out rod. The setting out rod is particularly crucial when manufacturing curved or complex work as it gives the joiner a true image of how the completed component will look.

Setting out rods should include horizontal and vertical sections through the components as well as elevations of any complex areas.

Rod marked up for casement window

Remember

Although rods are marked up full size and to scale, it is good practice to mark on the sizes of the sections/components. This will save the joiner having to measure the sizes, saving time and effort.

Rods can be re-used by painting over them but it is a good idea to keep a record of the rods and store them for future reference.

Once the rods have been marked out, a cutting list can be made, the setting out rods can be stored to protect them from damage and the materials can be machined.

Now select the timber and cut to size, bearing in mind which will be the face and the edge. The face and edge are usually the two best adjacent sides, as they are most likely to be seen. Note the position and severity of any defects, as it may be possible to remove these when machining any rebates or grooves.

Next you must mark out the timber, which involves two main operations.

First you must apply face and edge marks, which serve as a reference point for the rest of the marking out and machining. The marks are shown right.

Now for the last stage in making out: the position of the joints, etc. To do this, you transfer the information from the setting out rods to the timber members. You should transfer the information as clearly as possible to avoid any confusion.

Now the joints can be cut and the component assembled.

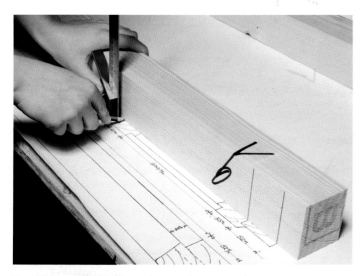

Transferring details from the setting out rod

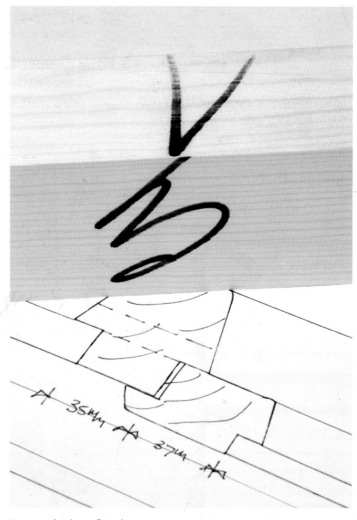

Face and edge of timber

Windows

Windows are expected to stand up to the elements, so they usually need to be made from a solid construction.

In this section, our example will be a simple casement window, 900 mm high and 500 mm wide.

Setting and marking out a window

First you need to make setting out rods drawn full size to show the sections of the height and width.

Did you know?

The method of setting out and marking is virtually the same for every component. Remember this when you look at the following examples of manufactured components.

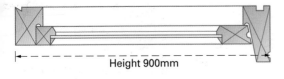

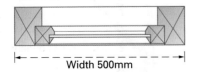

Figure 12.1 Height and width rods

Use this drawing to produce the cutting list, then use the cutting list to prepare the materials for marking out.

When manufacturing a window, two different framing methods are used: one for the fixed parts and another for the moving parts. In our example, the fixed part is the window frame, and the moving part is the opening sash. For the frame, the horizontal members (head, sill) are mortised and the vertical members (jambs) are tenoned; for the moving sash, the opposite is true – the horizontal members are tenoned and the vertical members mortised.

Activity

Using dimensions dictated by your trainer, produce setting out rods for a window.

Activity

Produce a cutting list for the window in the previous activity.

Remember

Don't forget that opposite members such as the head and sill or jambs should always be marked out in pairs.

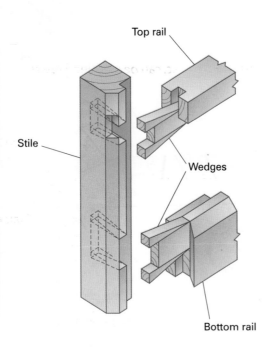

(a) Sash construction **(b)** Frame construction

Figure 12.2 Joints in frames and sashes

Once the materials are machined, you can mark them out by transferring the marks from the setting out rod to the various members.

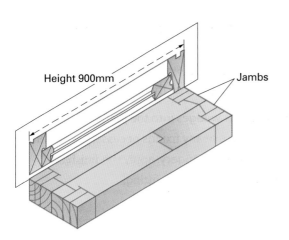

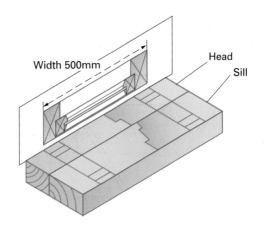

Figure 12.3 Members marked out

Once all members are marked out, the joints, rebates, grooves, etc. can be machined ready for assembly.

Assembling a window

The first step in assembly is known as dry assembly. This is where the window is assembled without adhesive to check that the frame will be square and the joints will fit tightly. If the frame needs adjustment, this is the time to do it. After any adjustments, the frame can be glued up and clamped.

As soon as the frame is clamped but *before* the adhesive sets, you need to make two important checks.

First you must check for square, in one of three ways:

- **Use a large square** – simply try the square in all four corners to check for square (only in small frames).

- **Use the '3, 4, 5' method** – measure horizontally from one corner 300 mm, then measure vertically down from the same corner 400 mm: the distance between these two marks should be 500 mm.

- **Measure the diagonals** – either simply measure from one corner to the other and check that the measurement is the same on the other diagonal, or use a **squaring rod**.

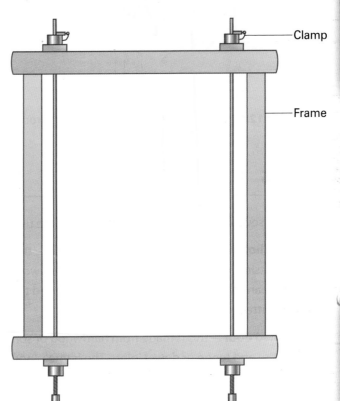

Figure 12.4 Clamping up a frame

Remember

It is good practice to remove any marks or pencil lines *prior* to assembly, as they can be difficult to remove once the component is fully assembled.

Activity

Using the setting out rods drawn in the previous activity, transfer marks to materials ready for machining.

Definition

Squaring rod – a piece of timber used to check the diagonals. The squaring rod is laid across the diagonals and marked at each corner, then moved to the other diagonal to check that the marks line up

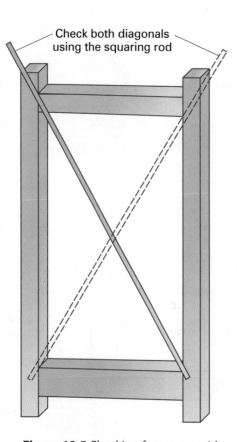

Figure 12.5 Checking for square with a squaring rod

The second check you must make is for **winding**. The simplest way to do this is to ensure that the frame is laid on a flat surface and sight through the frame to check that there is no twist.

Now hammer wedges into the joints to tighten them up and use any fixings necessary. You can then fit any ironmongery, apply the finish and send the window out for fitting.

Doors

There are several types of door, but all doors come under one of two main categories.

- **Flush/hollow core door** – as the name implies, this is a hollow door with a frame around the outside, clad usually with hardboard or plywood. The interior of the door is packed with cardboard for strength, and a lock block is fitted to one of the stiles to allow a lock or latch to be fitted.

- **Framed door** – this is made from hardwood or softwood and constructed using either mortise and tenon or dowel joints. The frame is normally rebated to incorporate solid wood panels or glass, onto which beads will be fitted.

Flush doors and dowel-construction framed doors are normally factory-produced, so here we will look at what is involved in the production of a mortise and tenon framed door.

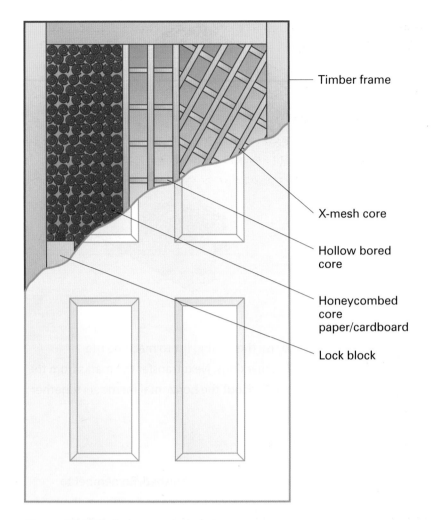

Figure 12.6 Flush door exploded view

Timber frame

X-mesh core

Hollow bored core

Honeycombed core paper/cardboard

Lock block

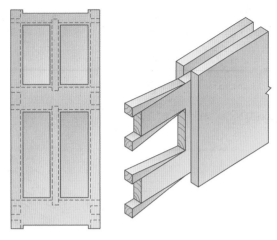

Figure 12.7 Mortise and tenon construction

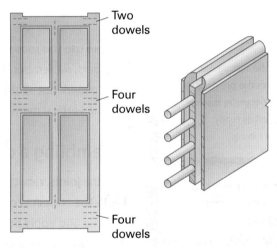

Two dowels

Four dowels

Four dowels

Figure 12.8 Dowel construction

Setting out and marking a door

As always, you start with setting out rods showing a section through the height and width.

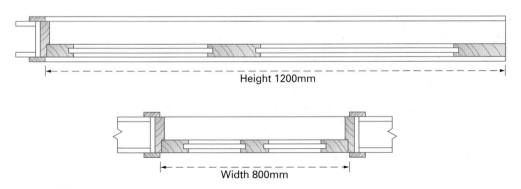

Height 1200mm

Width 800mm

Figure 12.9 Height and width rods

Activity

Using dimensions dictated by your trainer, produce setting out rods for a door.

Activity

Produce a cutting list for the door in the previous activity.

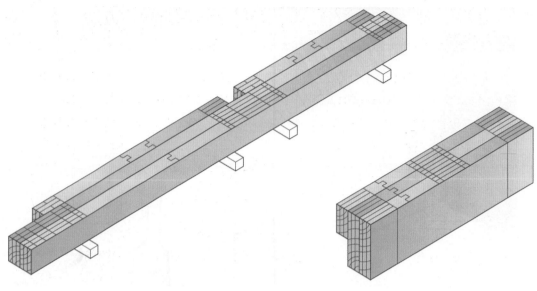

Figure 12.10 Members marked out

You can use the rods to produce a cutting list, and use the cutting list to machine the members, remembering to apply the face and edge markings. Next transfer the marks from the rods to mark out all the members, remembering to mark out the horizontal members together and the vertical members together.

Assembling a door

Now the joints, rebates, etc. can be machined and the door can be finished. Remember to follow the correct sequence: dry fit, glue up, clamp up, check for square and winding, then clean up and apply finish. The door is now ready to be hung.

Activity

Using the setting out rods drawn in the previous activity, transfer marks to materials ready for machining.

Activity

Machine and assemble the door set out previously.

Knowledge refresher

1   What drawings should setting out rods contain?

2   Why should setting out rods be stored after the component is made?

3   What should be the first stage in assembly and why?

4   State two ways of checking a frame for square.

5   Name two main categories of door.

What would you do?

You have been tasked with making a series of windows. You draw out the rods, transfer the marks and machine the timber. You then glue and cramp up the frames and as it is the end of the day you leave them overnight to dry. When you come back in the next day you remove the frames and find that they are out of square, some joints are poor fitting and the frames are also twisted. What has caused these problems? What needs to be done to remedy these problems? What effect will these problems have on productivity and profit? What should have been done to prevent the problems from occurring?

Stairs

Stairs are set out and marked out differently from doors and windows. To set out stairs, you first need all the dimensions, including the rise, going, etc.

Two templates are needed to mark out the tread and riser positions on the strings: the pitch board and the tread and riser template.

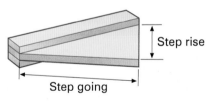

Figure 12.11 Pitch board

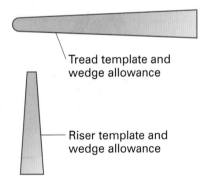

Figure 12.12 Tread and riser template

Remember

There are various regulations concerning stairs, and all dimensions must comply with these regulations. See Chapter 7 pages 184–5 for details.

Setting out and marking stairs

First machine the strings to the required sizes and set them out on the bench, remembering to mark the face and edge marks.

Mark the pitch line using a margin template for accuracy.

Remember

Take time and care when making templates so that they are as accurate as possible.

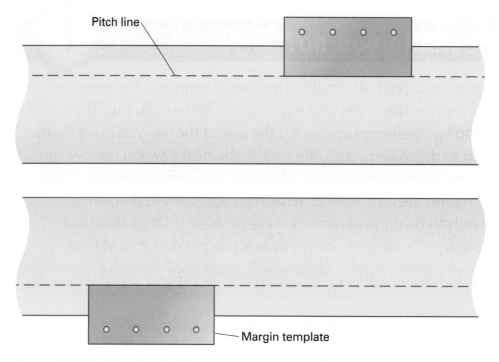

Figure 12.13 Marking the pitch line using a margin template

Set a pair of dividers to the hypotenuse of the pitch board, then mark this distance all the way along both the strings: the two points of intersection will establish the tread and riser points relevant to the pitch line.

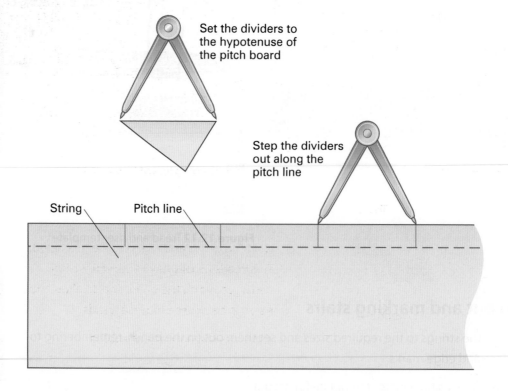

Figure 12.14 Marking the tread and riser points

The pitch board can now be used to mark the rise and going onto the strings and the riser and tread templates can be used to mark out the actual position of each step.

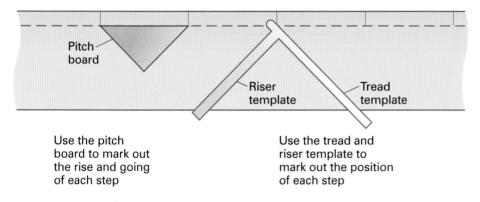

Use the pitch board to mark out the rise and going of each step

Use the tread and riser template to mark out the position of each step

Figure 12.15 Marking the rise and going, and the position of the treads

Now router out the housings – it is best to use a stair housing jig combined with a router for this job.

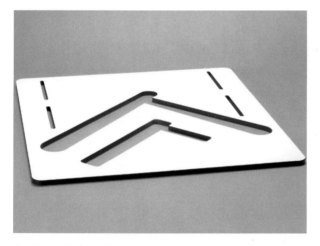

Router stair-housing jig

Stair-housing jig used with router

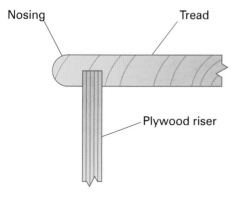

Figure 12.16 Joint between tread and riser

Next the treads and risers can be machined. The treads usually have a curved nosing and a groove on the underside, to allow the risers to be housed into the treads.

The stairs can now be assembled.

Assembling stairs

Start by laying one string on the bench with the housings facing upwards. Put in place all the treads and risers, ensuring that the tread/riser ends and the housing are glued. Place the other string on top of the strings and risers, glue, then apply clamps.

Tread and riser assembly

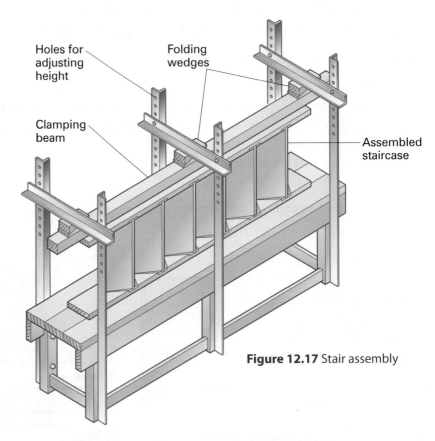

Holes for adjusting height

Folding wedges

Clamping beam

Assembled staircase

Figure 12.17 Stair assembly

Drive glued wedges into the housings to force the treads and riser to the front of the housing, making sure there are no gaps.

Once all the wedges are in place, check the front of the staircase to ensure that the treads and risers are situated perfectly. Now screw the bottom of the risers to the back edge of the treads. Finally fit glue blocks where the tread meets the riser, at the internal angle.

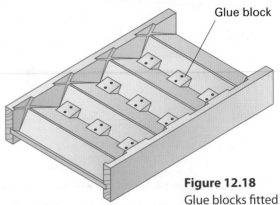

Glue block

Figure 12.18
Glue blocks fitted

Once the adhesive has dried, remove the clamps, then clean up the staircase, protect it with hardboard or bubble wrap and send it on site to be fitted.

Basic units

The most common type of unit you will deal with is the kitchen unit: if you understand the principle of assembling these, you will be able to assemble any unit.

Most units are made from melamine-faced chipboard as this is light and easy to wipe clean, but other materials such as MDF, plywood and even solid timber can be used.

There are two main methods of unit construction.

- **Box construction** – also known as slab construction, this comprises vertical standards, rails and shelves, often with a plinth and bottom rail attached, with the remainder of the shelves sitting on adjustable brackets. With this method, the back panel is fixed to hold the frame square and make the unit sturdier.

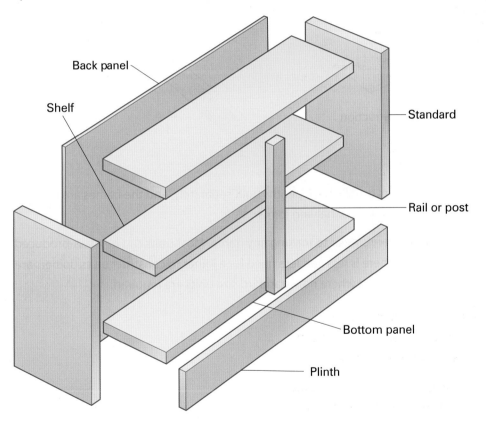

Figure 12.19 Box construction

Activity

Using dimensions provided by your trainer, mark out and machine the components required to create a framed construction unit.

- **Framed construction** – also known as skeleton construction, this comprises either a pair of frames joined together with rails, or cross frames joined together by rails at the front and back.

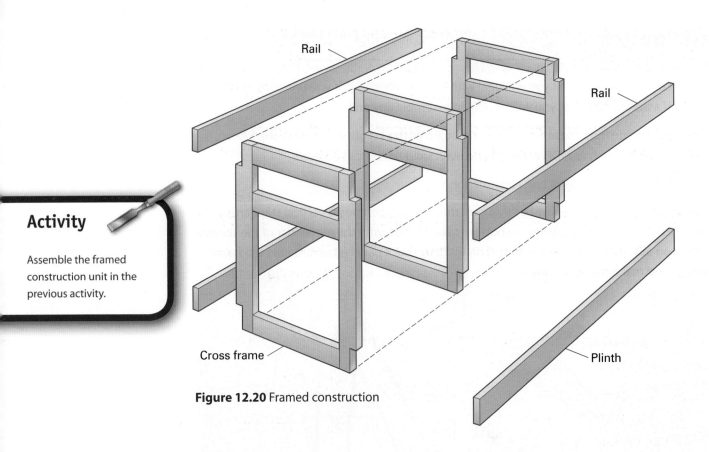

Figure 12.20 Framed construction

Framed construction units usually have plinths built separately, and the frames are jointed using either mortise and tenon or halving joints.

Making these frames is very similar to making any frame. Most units can be mass-produced and bought cheaply, so there is less and less need for a joiner to construct units. Joiners are mainly used for specialist jobs where non-standard size units are required.

FAQ

Do face and edge marks have to be used?

No, they don't, but without them there is a good chance that mistakes will be made. Using face and edge marks is considered to be very good practice.

Which type of joint is best for a door: the mortise and tenon or dowel?

The mortise and tenon is the more traditional way, but the dowel is far easier. Both joints work well and the choice of which to use is down to the joiner or client.

Why are glue blocks fitted in stairs instead of just screwing down through the face of the tread?

Glue blocks prevent any movement once the stair is in use. They are preferred to screwing as screwing is unsightly and there is a risk of splitting the nosing.

Knowledge refresher

1. What is the best way to cut out the housings on a stair string?

2. What is the purpose of glue wedges?

3. State four materials from which units can be made.

What would you do?

You have been tasked with manufacturing a straight flight staircase. You mark out the strings individually and then machine the treads, risers, etc. You then router out the strings and prepare the staircase for assembly. Once the frame is assembled, you see that the stair seems to be out of square: one string projects past the other by at least 2", so when the staircase is laid flat, the treads are well out of level. What could have caused this? What can be done to rectify it? What could have been done to prevent it?

chapter 13

Mark, set out and manufacture complex joinery products

OVERVIEW

With the basics of component manufacture covered in the previous chapter, we will now look at more complex components.

This chapter will cover the marking, setting out and manufacture of:

- louvre ventilator frames
- complex doors with shaped heads
- geometrical stairs
- laminated components.

Louvre ventilator frames

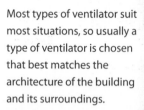
Louvre ventilator frames are commonly found in areas that require constant ventilation, such as boiler rooms. The ventilator is made up of a frame housing louvre boards, pitched at either 30, 45 or 60 degrees. Louvre frame ventilators are usually rectangular or square but for the purpose of this book we will look at shaped louvre ventilator frames. These are most commonly pitched, circular-headed or gothic.

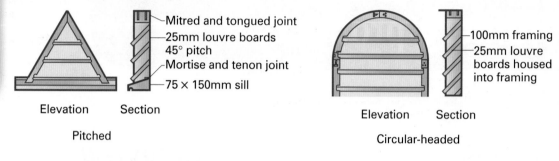

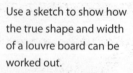
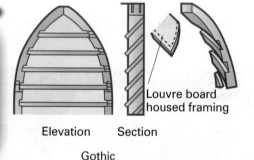

Figure 13.1 Pitched, circular-headed and gothic louvre ventilators

The first style we will look at is the pitched ventilator.

Pitched ventilator frame

The first step, as always, is to draw out the frame full size. From this drawing we can get the true width and shape of the louvre boards.

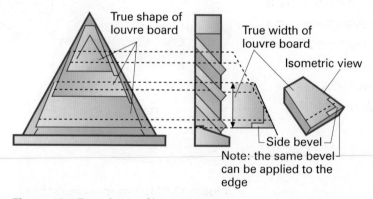

Figure 13.2 True shape of louvre boards

The louvres can then be marked out and cut, and the stock material for the frame can be machined.

The next stage is to mark out the frame joints and the housings, again using the full-scale drawing. The first step is to project the position of the housings from the section (A) onto the elevation (B). Machined stock (C) can be laid onto the drawing and the housings can be transferred from the elevation (B). The members should be marked out in pairs to ensure accuracy and that the finished frame is square.

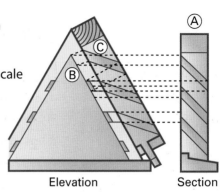

Elevation   Section

Figure 13.3 Marking out for housings

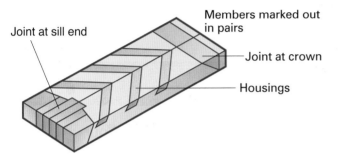

Figure 13.4 Marking out

The frame is now ready for assembly and finish.

Circular or curved ventilator frame

Circular or curved frames are more complex and the method of producing curved members is explained more fully in the next section.

The setting out process for circular headed and gothic frames is broadly the same as for the pitch frame, but the marking of the louvre positions is more difficult. It is best to use a square frame and a straight edge to mark the positions of the louvres on the face, and then use a pitched block to mark out the inner side of the frame.

> **Activity**
>
> Using the dimensions and materials provided by your trainer, mark out and machine the materials required for a gothic-head ventilator frame.

> **Activity**
>
> Assemble the gothic-head ventilator frame you prepared in the previous activity.

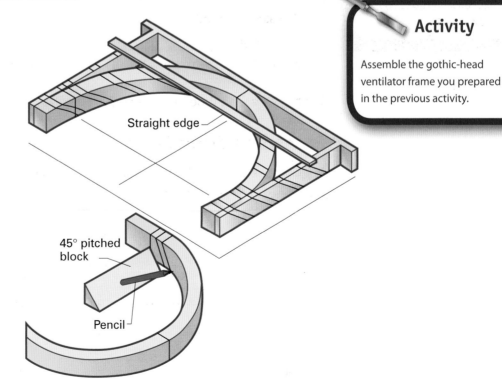

Figure 13.5 Marking for a circular head

Knowledge refresher

1 What are the three angles that louvres are normally pitched at?

2 Why are louvre ventilation frames necessary?

What would you do?

You have been tasked with creating a pitched louvre frame. Your employer gives you the overall dimensions of the opening. What other information do you need?

Complex doors with shaped heads

Doors with shaped heads can provide a sense of grandeur but have to be made to measure, which can prove costly in today's market where most doors are mass-produced.

There are four main types of shaped head door:

- segmental
- semicircular
- gothic
- parabolic.

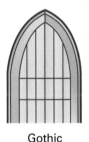

| Segmental | Semicircular | Gothic | Parabolic |

Figure 13.6 The four main types of shaped head doors

When working with shaped head doors, many different types of joint can be used to create curved or arched members, for example, hammer headed keys, tenon joints or joints secured using handrail bolts.

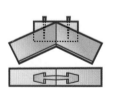
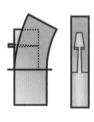
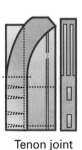

| Joint at crown | Joint at springing line | Tenon joint | Handrail bolt joint |

Figure 13.7 Hammer headed keys, tenon joints and handrail bolt joints

Basic method for producing a shaped head door

The example we use here is a semicircular door but whichever type of door is to be made, and whichever joints are used, the method is the same.

Working shaped members involves similar operations to working straight members, though greater care and skill are needed to maintain a uniform standard of craftsmanship. Vertical spindle moulders and routers are the best tools to use for creating the curved sections, and accurately made face templates are vital for cleaning and shaping the curved members. The height and width rods should be made as normal, but the circular head section must be marked out full size.

To mark out the shaped head, begin with the centre line and the springing line, where the joints at the stiles will be. Next mark on the stiles, and finally mark out the curved head with either a beam compass or radius rod.

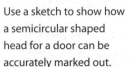

Activity

Use a sketch to show how a semicircular shaped head for a door can be accurately marked out.

Activity

Use a sketch to show the different joints that can be used to join the curved members on a door to other curved or straight members.

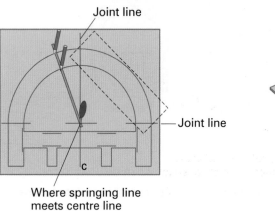

Figure 13.8 Setting out for a circular head

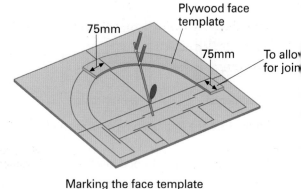

Figure 13.9 Face template

Next mark the face template from the drawing, making the template a minimum of 75 mm longer at each end to allow for the joints to be constructed. The face template must be made from plywood at least 9 mm thick.

Now machine the materials. The curved sections can either be formed on the vertical spindle moulder, with the face template used as a jig, or cut roughly on the band saw and tidied up using the face template and a router. Once the material has been planed and cut to the correct dimension, mark it out, remembering to mark out the stiles in pairs and the bottom, middle and frieze rail together.

The joints between the stiles and the rails are all mortise and tenon, so they

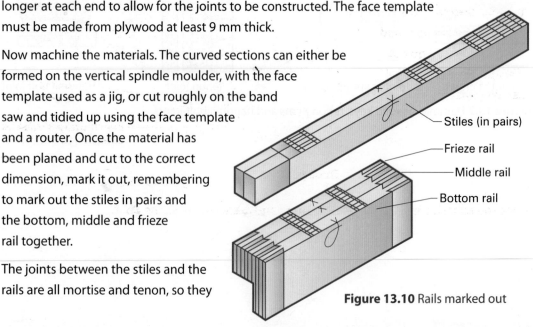

Figure 13.10 Rails marked out

can be manufactured and dry fitted to ensure a good tight fit. The joint between the stile and the circular head is a twin mortise and tenon – see Figure 13.11 for the detail.

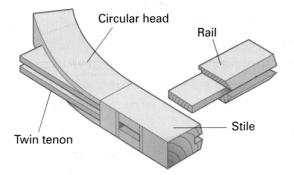

Figure 13.11 Twin tenon and rail

Activity

Using dimensions and materials stated by your trainer, measure, mark out and machine the timber for a semicircular headed door.

Joint the crown using handrail bolts with cross tongues for additional strength, as in Figure 13.12:

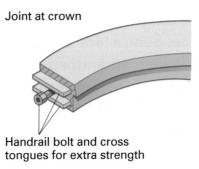

Figure 13.12 Crown joint

Activity

Assemble the door you prepared in the previous activity, using any suitable joints.

Once all the joints have been machined and tested for dry fitting, the door can be glued, assembled and clamped. When clamping the door, take care to ensure that the joints remain tight and the door does not go out of shape. Here is the ideal clamping method:

Figure 13.13 Door clamped in place

Once the adhesive is set, remove the clamps, and glaze and finish the door.

Knowledge refresher

1 Name three types of shaped head doors.

2 State three jointing methods used when creating shaped head doors.

3 Why does a face template need to be accurate?

What would you do?

You have been tasked with creating a semicircular headed door. You mark out all the components and machine them. When you tidy up the curved pieces using a face template and router, there are several bumps and hollows in the finished piece. What could have caused this? What needs to be done to remedy this? What should have been done to prevent it?

Geometrical stairs

Stairs are said to be geometrical when they have continuous strings and handrails. Geometric stairs are very complex, and building them requires a wide understanding of geometry. In this section you will find a brief overview of the geometry, and will see how to construct a basic geometrical stair.

Producing a geometrical stair

Our example here is a geometrical stair with a quarter over a **wreathed string** with a quarter-turn of winders (though this sounds complex, it is essentially a quarter-turn stair).

Notice that there is no newel post at the turn; instead the outer string is a continuous string, better known as a wreathed string. The wreathed string will be shown in more detail later but first we will concentrate on the wall strings.

Definition

Wreathed string – continuous string that rises while turning

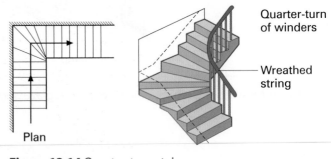

Plan

Quarter-turn of winders

Wreathed string

Figure 13.14 Quarter-turn stair

The first thing, as always, is to do a drawing so that you can work out the size of the winders, etc. You must remember to keep within these regulations:

- the rise of the tapered steps must be the same as the rise of the other steps
- the tapered step must not be less than 50 mm at the narrowest point
- the going of the tapered steps (measured at the centre of the steps) must be the same as the going of the other steps.

The stairs rise as they turn, so the wall strings need to be made wider. There are two ways to make the wall string: one is to make the string out of wider stock and cut away the waste, but this is very costly; the preferred, cheaper method is to attach pieces to the wall strings. The attached pieces should ideally be **biscuit-jointed** and glued for strength, though when the treads and risers are fixed to the string this will strengthen the joint.

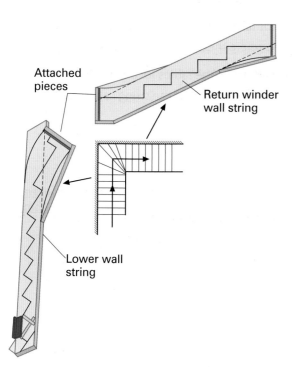

Figure 13.15 shows why the attached pieces are needed, and how the wall string should be marked out for the treads and risers. To maintain accuracy, you should take the sizes for marking out the winders, etc. from the drawings.

The housings for the risers and treads, and the tongue and grooved joint where the two wall strings meet, can all be machined.

Figure 13.15 Wall strings

Definition

Biscuit joint – a type of tongued and grooved joint that works like a dowel joint, but instead of a dowel a flat oval plate (biscuit) fits into a slot. When PVA adhesive is introduced, the biscuit expands and makes a very strong bond

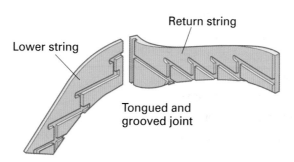

Figure 13.16 Joint between wall strings

Activity

Using measurements and materials provided by your trainer, measure out and mark the wall strings for a quarter-turn stair.

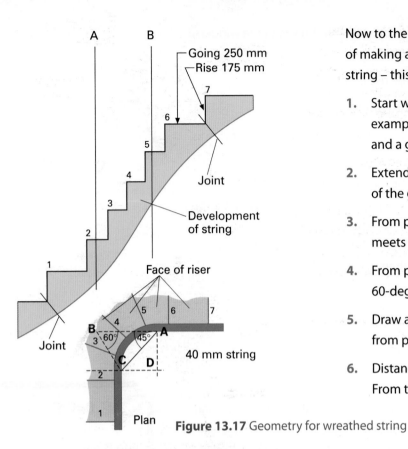

Figure 13.17 Geometry for wreathed string

Now to the wreathed string. Probably the most difficult aspect of making a wreathed string is finding the curved shape of the string – this is where a good grasp of geometry is required!

1. Start with a full-scale plan drawing of the stair (for this example we will use a 40 mm string with a rise of 175 mm and a going of 250 mm).

2. Extend a horizontal straight line from point A the distance of the going (250 mm) creating line A–B.

3. From point B draw a line down at 60 degrees to where it meets the outside of the string.

4. From point A draw a line down at 45 degrees to meet the 60-degree line, forming point C.

5. Draw a horizontal line from point C and a vertical line down from point A: where these two lines meet is point D.

6. Distance A–D will give the radius for a going of 250 mm. From this the development of the string can be created.

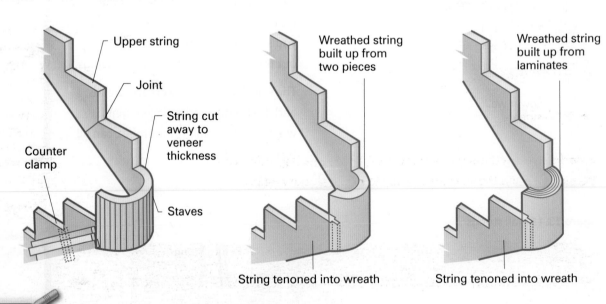

Figure 13.18 Wreathed string construction

The wreathed string is normally a cut string (as opposed to being routed) and consists of three parts. The wreathed part is the curved part, joined to a top and bottom cut string. This join can be done in three different ways, the most common of which is to use staves, which we will look at here (laminating is covered later in the chapter).

The string starts out as a normal straight string piece. First you must reduce the string's thickness at the curved area to a veneer thickness of around 3 mm. Once the section has been removed, place the string over a drum made to the shape of the curve and fix it in place, taking great care when bending the string. Glue a series of tapered staves into place and allow them to set.

If the string is going to be seen, and a quality finish is required, fit another veneer to the string to hide the staves.

Once the adhesive has dried, remove the string from the drum and joint it to the top and bottom cut string. The joint used depends on the thickness of the string and how the curve is formed. Three of these are familiar to you already – bridle, halving and mortise and tenon joints – so here we will look at the counter clamp method.

The counter clamp method is a simple but effective way of jointing the strings, working along similar lines to draw boring:

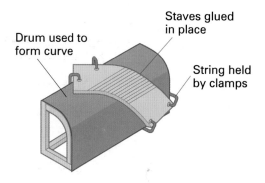

Figure 13.19 String over drum

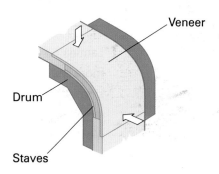

Figure 13.20 Veneer being fitted over staves

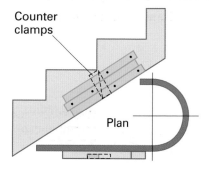

Figure 13.21 Counter clamps

Once the wreathed strings are completed, fix the treads, winders and risers in place and the staircase is ready to be fitted and finished.

Activity

Using measurements and materials provide by your trainer, create a wreathed string using a suitable method.

Knowledge refresher

1   Stairs are said to be geometrical when they have what?

2   State one extra regulation that is required when making a staircase with winders.

3   Why are the staves that are fitted into the wreathed string tapered?

What would you do?

You have been tasked with creating a wreathed string using a drum method. You reduce the string's thickness and then attempt to bend it over the drum. The string snaps, so you try again, and it snaps again. What could be causing this? What needs to be done to rectify it?

Laminated components

Laminated components are widespread throughout the construction industry, and are used more and more nowadays, the most basic form of laminate being plywood. Laminates are simply thin strips of timber that are glued together. In carcassing, particularly for roofing, laminates have an advantage over solid timber as laminated beams can be made to any length. Most laminated beams are now mass-produced and machined with specialist machines, but for the purpose of this book we will look at traditional methods for laminating a curved piece and a long beam.

Laminating a curved piece

Laminated curved pieces can be used for almost any purpose, and can be made to any radius, or even a variety of radii on a single piece to create a serpentine effect. Laminated timber can even be used for the shaped door heads discussed earlier in this chapter.

The amount of laminate needed depends on the thickness required, which in turn depends on the curve. For our example, the finished timber will be 60×60 mm with a curve that is not too severe, so we will be using 3 mm thick laminate.

Commonsense might say that 20×3 mm laminates will give you a 60 mm finish, but this will probably not be the case as there will be a thin layer of adhesive between each laminate, increasing the finished size to more than 60 mm. Nineteen laminates may be enough, but this may leave you slightly short of the 60 mm. One option, depending on the precision of the finish required, might be to use 20 laminates, then plane the laminated component to the exact size using a spoke shave plane.

It is good practice to machine more laminates than you will use as this allows for any strips with shakes or dead knots, and any damage that might occur in the manufacturing process. The 3 mm laminates should be sawn and ideally planed to the exact thickness. Ensure that all the laminates are uniform and clean, to create a good bond.

Activity

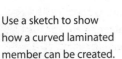

Use a sketch to show how a curved laminated member can be created.

Once the timber is prepared to the correct size, it is ready for shaping. Shaping can be achieved by:

- **dry clamping** – the timber is placed into the jig and forced into shape by the clamp. This method yields poor results as the timber needs to be kept in the clamps for a long time, with no guarantee that it will keep its shape.

- **wet clamping** –the timber is soaked, then clamped in place. This produces better results than dry clamping, but again the timber must be kept clamped until it has dried out, which can be time-consuming. Also, with this method there is a good chance that the timber will spring back and not meet the required radius.

- **steam clamping** – the timber is placed into a steam box for a set time, then clamped into place. This is by far the best method: it does not take as long and gives the best results.

Timber is placed into the steam box and steam is pumped in. The heat allows the timber to be bent, while the moisture stops the timber becoming brittle and snapping. The length of time that the timber should be in the box depends on the type and thickness of the timber and the severity of the bend required, but 1 hour per 6 mm thickness is a good guide time (it is better to oversteam the timber rather than understeam). Timbers should be placed into the box with piling sticks between them to allow the steam to circulate. After the required amount of time the timber should be removed.

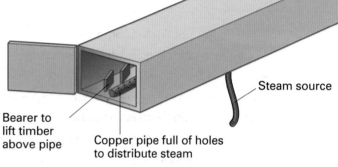

Steam source

Bearer to lift timber above pipe

Copper pipe full of holes to distribute steam

Figure 13.22 Steam box

The clamping of the timber can be done in several ways, but the best way by far is to make up a jig.

This jig is an example only: any type of jig can be used.

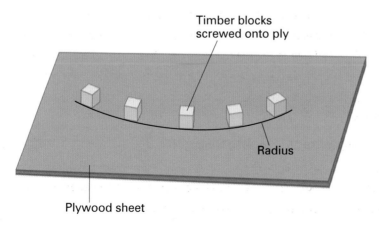

Timber blocks screwed onto ply

Radius

Plywood sheet

Figure 13.23 Typical jig

Safety tip

When using a steam box, always wear the correct PPE as the timber will be hot and placed immediately into the jig and clamped in place.

Did you know?

A steam box must be manufactured from a suitable material such as WBP plywood – otherwise the timber making up the box will break down from the effects of the steam!

Activity

Using information and materials provided by your trainer, create a jig used for creating curved laminated components.

With modern advances in adhesives it is now possible to glue up the timber straight out of the box and clamp it into the jig, but traditionally the timber is placed into the jig dry (i.e. not glued) and left overnight to dry out, then glued and clamped the next day.

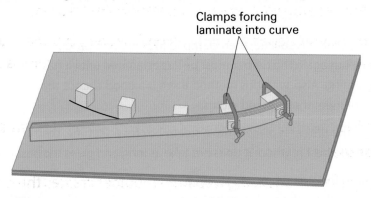

Clamps forcing laminate into curve

Figure 13.24 Timber clamped into a jig

Use the adhesive manufacturer's guidelines to decide the length of stay in the jig. Once dry, the laminated component can be cleaned up ready to be fitted. There may be a slight spring back once the clamps are released. This can be overcome by making the radius on the jig slightly more than required: when the spring back occurs it will leave the laminate at the correct radius.

Laminating a long beam

Laminating a long beam is a simple process with no need for steaming (unless the beam has a bend in it). A laminated beam can be made to any length because the laminates can be staggered.

The procedure starts with the required number of laminates planed to the required thickness, which for this example is 50 laminates at 5mm. The laminates are then glued together and clamped.

Staggered joints

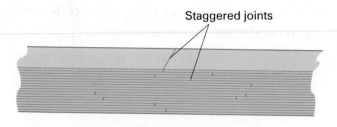

Figure 13.25 Staggered laminates

FAQ

How many louvres should there be in a ventilation frame?

The number of louvres depends on the client's wishes, but the spacing is very important, as they should not allow birds or other animals into the building.

Which is the best way to create a shaped head: using a band saw and router or using a spindle moulder?

Both ways are suitable, but using a spindle moulder creates the best finish and is quickest as it can be set to create moulded curved pieces in a single pass.

On a geometrical stair, does a wreathed string need to be a cut string?

No. The treads can be housed into the string, but with the curve this is very difficult so it is easier to have a cut string.

Do I need to cut veneers for both sides of the wreathed string?

This depends on the finish required. If the string is visible and has a varnish or stain finish, it is good practice to have veneers on both sides, but a painted string does not need veneers.

When laminating, what is spring back?

Spring back occurs when the clamps are released and the timber springs back into its natural state.

Knowledge refresher

1. Give a reason why laminated beams are preferred to solid timber.

2. What is a suitable guide time for steaming laminates?

What would you do?

You have been asked to create a curved laminated component that will be used in the head of a doorframe. Your employer tells you to cut up some laminates and put them in the steam box for about an hour. Do you have all the information you need? What other information do you need? What could the outcomes be if you don't find out the extra information?

Glossary

abbreviations	shortened versions of words or names, often using just the initial letters of each word
acoustic	to do with the way sound travels around
amenities	facilities such as toilets, rest areas, etc.
asbestosis	a serious lung condition caused by breathing in asbestos
ballistic tool	any tool that propels an object, e.g. a nail gun
balustrade	unit comprising handrail, newels and the infill between it and the string, which provides a barrier for the open side of the stair
barrier cream	a protective cream that stops water and other substances getting through
beam compass	a special carpenter's compass for creating arcs and circles with a wide diameter
blemish	a stain, scar or spot
'bottleneck' effect	when things get jammed, especially when a large flow comes into a narrow passage and not all of it can get through, as in the neck of a bottle
clout nail	a large-headed nail
combustion	burning, the action of fire
compliance	obeying, fitting in with
conservation	preservation of the environment and wildlife, or of rare or special buildings
contamination	when harmful chemicals or substances pollute something (e.g. water)
contingencies	plans set up just in case something happens
corrosive	a substance that can damage things it comes into contact with (e.g. material, skin)
counterbalance	a weight used to 'counter' or act against the weight of something else, so that it doesn't fall over

cut-outs	scaled-down shapes that represent bigger things, which can be used on a scale drawing to help you plan
decibel (dB)	the standard unit for measuring noise level
dermatitis	a skin condition where the affected area is red, itchy and sore
discrepancies	when there is a difference or variation between two things that should be the same
dismantle	take apart, take down carefully
DPC	damp proof course, a substance that is used to prevent damp from penetrating a building
disproportionate	out of proportion to something else, far more or less than you would expect
duration	how long something goes on
electrocution	death or injury through coming into contact with an electric current
employer	the person or company someone works for
employee	the person employed by the employer, the member of staff
exhaustive	absolutely complete
going	the depth of a step (the measurement from a step's riser to the edge of the step)
halving joint	the same amount is removed from each piece of timber so that when fixed together the joint is the same thickness as the uncut timber
Health and Safety Executive (HSE)	government organisation that enforces health and safety law in the UK
hydraulic	worked by water or other liquid in pipes
hypotenuse	the side of a right-angled triangle opposite the right angle
impregnated	soaked right through

inconsistencies	when things are not the same, not consistent	**sash**	part of an old-fashioned style of window with a top and a bottom section that can be moved up and down with cords
in situ	on the spot, in the actual location where it is needed		
institutional	to do with an institution such as a hospital or school	**segmental**	made or divided into sections, usually equal sections
interim	in the time between, for the time being, as a holding measure	**serpentine**	like a snake, curving one way and the other
intersection	the point at which things join or cut across each other	**serrated**	with teeth on, like a saw
jeopardise	endanger, put at risk	**silicone**	a chemical used in rubbers, seals and polishes that is waterproof, very stable and long lasting
legislation	laws or the making of laws		
manifestation	how something looks or is presented		
multiples	sets of more than one; in housing, sets of matching dwellings that are all the same	**skew-nailed**	nailed with the nails at an angle
		solvent	a substance that dissolves another e.g. paint stripper
node	a key point (especially on a critical path chart)	**stipulation**	a condition of an agreement, a particular term of a contract
nogging	a short length of timber, most often found fixed in a timber frame as a brace	**string**	main board to which treads and risers are fixed
		superseded	overtaken by, replaced by
no-show	when someone does not turn up as planned	**surveillance**	carefully watching over or keeping an eye on
objectives	aims, purposes	**susceptible**	prone to, vulnerable to
out of square	not properly at right-angles, a bit skew	**sustainability**	the ability to last or carry on, how easy something is to keep going
paramount	the most important thing, the greatest concern, above all others	**tapered**	getting thinner towards one end
penalty clause	a clause in a contract saying a fine has to be paid, or some other penalty made, if a certain thing happens, e.g. the job overruns	**telescopic**	sliding or arranged like the joints of a telescope, so it can be packed away small
		toxic	poisonous
prevalent	common, often found	**tusk tenon joint**	a kind of mortice and tenon joint that uses a wedge shaped key to hold the joint together
prohibition	a ban, saying something cannot happen or be done		
projection	sticking out, a part that juts out	**uPVC**	unplasticised polyvinyl chloride, a relatively stiff material, sometimes brittle in cold weather, of which sewer pipes or water pipes are made.
prosecute	take someone to court for committing a crime		
prospective	likely or possible in the future, but not actually happening or approved now	**vacuum**	a space that is completely empty, with all the air sucked out
radius rod	a rod which acts as a large compass	**ventilation**	to do with air flow
rectify	put right	**vibration white finger**	a condition that can be caused by using vibrating machinery (usually for very long periods of time). The blood supply to the fingers is reduced which causes pain, tingling and sometimes spasms (shaking)
recurrence	happening again		
remit	scope, job, the areas an organisation or individual has to cover		
residential	where people live, rather than a business district, for example		
reverberation	echoing, reflecting sound and vibration	**vice versa**	the other way round
		WBP	water/weather boil proof
roofing ready reckoner	a set of mathematical tables giving a quick way to work out rafter lengths		

Index